CAMBRIDGE COMPARATIVE PHYSIOLOGY

GENERAL EDITORS:

J. BARCROFT, C.B.E , M.A., F.R.S.

Fellow of King's College and Professor of
Physiology in the University of Cambridge

and

J. T. SAUNDERS, M.A.

Fellow of Christ's College and Demonstrator
in Animal Morphology in the University of
Cambridge

FEATURES IN THE ARCHITECTURE OF PHYSIOLOGICAL FUNCTION

FEATURES IN THE ARCHITECTURE
OF PHYSIOLOGICAL FUNCTION

BY

JOSEPH BARCROFT

C.B.E., M.A., F.R.S.

Fellow of King's College, Cambridge

Hon. M.D. Louvain, D.Sc. Queen's and National
Universities of Ireland

CAMBRIDGE

AT THE UNIVERSITY PRESS

1934

CAMBRIDGE
UNIVERSITY PRESS

University Printing House, Cambridge CB2 8BS, United Kingdom

Cambridge University Press is part of the University of Cambridge.

It furthers the University's mission by disseminating knowledge in the pursuit of
education, learning and research at the highest international levels of excellence.

www.cambridge.org
Information on this title: www.cambridge.org/9781107502475

First published 1934
First paperback edition 2015

A catalogue record for this publication is available from the British Library

ISBN 978-1-107-50247-5 Paperback

To

MY WIFE

CONTENTS

CONTENTS

PREFACE

It must have been about the time of the Harvey Centenary that Sir John Rose Bradford delivered an address to the Cambridge Medical Society; in it he said, so nearly as I can remember, "The difference between physiology as taught now and in my youth is that now the student is given principles: then he was only given facts". I commenced to wonder what these principles were, and the list, as it occurred to me, was much the same as the headings of the chapters of this book.

Having been privileged to give the Dunham Lectures at Harvard in 1929, some of the titles became expanded into lectures, the publication of which could scarcely be refused when the request was made by the revered Founder of that Benefaction. The whole grew, supplying material for other lectures, and then a curious thing happened. At the outset I had regarded the body as a noble building on the principles which it exhibits as unconnected features in its architecture. It became clear that the features were far from being independent. The highest functions of the nervous system demand a quite special constancy in the composition of its intimate environment. The stability of the internal *milieu* almost compels the principle of the storage of materials and of integration in adaptation. Again an easy stepping stone to integration in the practice of the body to have more than one way of doing many things. But parallel mechanisms may express themselves not only in integrative but in antagonistic processes. Moreover, increased function actively may be achieved either by heightening the efforts of units already functioning or by marshalling a greater number of units: and so we arrive at the "all-or-none" relation.

It seemed almost as though there had emerged an approach

to physiology from an unusual angle: not from that of mere structure, whether the structure of organs or of chemical formulae, but from the principles of function.

It is doubtful whether the book would ever have seen the light but for the sympathetic encouragement of some of my friends, notably Sir Charles Sherrington and Prof. S. P. Cathcart. The help which I have received from Prof. Adrian, Mr Matthews, Dr Winton, Dr Keys and a host of other workers in the Cambridge Laboratory will be evident to all readers, and Mr Thacker has done much on the editing side. For the use of diagrams I am indebted to: Messrs Baillière, Tindall & Cox (*Veterinary Journal*); the Editors of the *Journal of Physiology*; the Council of the Royal Society, London (*Proceedings*); the Editor of the *Journal of Pathology and Bacteriology*; Messrs Walter de Gruyter & Co. (*Archiv für Anatomie u. Physiologie*).

J. B.

Cambridge 1934

CHAPTER I

"LA FIXITÉ DU MILIEU INTÉRIEUR EST LA CONDITION DE LA VIE LIBRE." (CLAUDE BERNARD)

INTRODUCTORY

Of the principles which govern the physiological processes of the human body, that of the fixity of its internal environment has been as thoroughly established as any. Within the last twenty years works of first-rate importance by Haldane, Henderson and Cannon have all dealt with the subject. Great progress has been made in our understanding both of the mechanisms which secure the constancy of the internal medium and of the exactness with which these mechanisms operate.

The principle enunciated by Claude Bernard, if dressed up in modern language, seems to me just a little grotesque. To say that the temperature of the body is adjusted to the tenth part of 1 per cent. on the absolute scale, or the hydrogen-ion concentration of the blood to the hundredth part of a pH so that the organism may obtain a free life, is surely to make a very ill-balanced statement. The accuracy of the first clause contrasts almost comically with the vagueness of the second.

Indeed this "up-to-date" version of Claude Bernard's statement constitutes almost a challenge. In general, the "end" is more important that the "means", therefore it seems at least desirable to make some effort towards arriving at a conception of the liberty of life to be attained by the fixity of the *milieu intérieur*.

There are two obvious avenues of approach to the problem.

It is instructive to study the efficiency attained in forms of life humbler than those endowed with a circulating medium of constant properties, to see how nature has ensured some degree of efficiency up to that point, ascertaining if possible the mechanism which ensures constancy and from what that mechanism is developed.

The second avenue of attack is that of breaking down the constancy of the internal "environment", preferably in that form of life—man—in which it is most highly developed, and noting the nature and degree of impairment which takes place.

To commence with, then, we may select certain physical and chemical properties of the blood which are maintained at an approximately constant value; we may endeavour to ascertain how this constancy was attained; we may then proceed to consider at what disadvantage, if any, the organism was, while as yet the internal circulating medium was variable.

Lastly, a word as regards the phrase "milieu intérieur": I do not regard the specialised hormones, such as adrenaline, as part of the general environment, nor do I include drugs. I have included sugar, possibly illogically, but it is very interesting.

HYDROGEN-ION CONCENTRATION

Every process of the living aerobic cell tends to alter the chemical composition of the medium in which the cell is situated in at least two ways. The mere fact of life tends to deprive the environment of oxygen and, if that environment be fluid, to increase its hydrogen-ion concentration. Not only is every cell in the body putting carbonic acid into the blood, but also the reaction of the circulating medium may be affected by special circumstances proper to the specialised activities of certain cells. Those of the pancreas when thrown into activity abstract alkali, and the oxyntic cells of the stomach abstract acid.

There is therefore every opportunity for the hydrogen-ion concentration of the internal environment to be inconstant. Yet in man it remains remarkably constant. The variation given in current books and taken from van Slyke's figure (1921) is $7 \cdot 0$–$7 \cdot 8$ pH, $i.e.$ roughly 1–5 gm. in 10^8 litres—about a fivefold variation. It is unnecessary to stress the smallness of the absolute quantity, it is equivalent to 1–5 gm. of hydrogen spread over the total volume of plasma of all the people in the United Kingdom, or about half the people in the United States. Yet that is the variation for the extreme limits of human life, the variation as between fatal coma and fatal convulsions. The concentration of hydrogen ions in the plasma of a healthy person is regulated about five times as exactly.

The data are given graphically (Fig. 1) by Arborelius and Lilje-strand (1923).

From rest up to work involving 50 litres per min. oxygen intake (945 kg. metres per min.), the alteration in hydrogen-ion concentration of the blood is given as from pH 7·33 ($cH = 4·8 \times 10^{-8}$) to pH 7·26 ($cH = 5·5 \times 10^{-8}$) for the average value obtained from determinations of the two authors.

Similar figures are given by Dill, Talbott and Edwards (1930) in Table I.

The amount of work done was of the same order. The extreme case was that of W. C., who did work corresponding to a total ventilation of 82 litres per min. The alteration in the hydrogen-ion concentration of his blood was from $cH = 3·6 \times 10^{-8}$ to $cH = 5·2 \times 10^{-8}$, a proportional increase of 1 : 1·45.

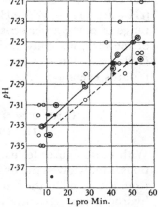

Fig. 1. Ordinate = reaction of blood, abscissa = ventilation in litres per minute; ○, individual observations on M. A.; ◉, average of observations on M. A.; ●, individual observations on G. L.; ◉, average of observations on G. L.

Table I

Subject	pH		Change of pH	Ventilation per kg. of body weight (litres)
	Rest	Work (running 20 min.)		
D. B. D.	7·42	7·39	−0·03	0·58
P. F. P.	7·44	7·32	−0·12	0·80
A. A. McC.	7·42	7·37	−0·05	0·72
J. H. T.	7·40	7·44	+0·04	0·67
W. C.	7·44	7·29	−0·15	1·19
O. S. L.	7·41	7·31	−0·10	0·67
J. L. S.	7·39	7·37	−0·02	0·72
H. T. E.	7·39	7·30	−0·09	0·59
A. V. B.	7·40	7·40	0·0	0·70
W. J. G.	7·36	7·38	+0·02	0·81
Average	7·41	7·36	−0·05	0·75

Instead then of a possible alteration of 500 per cent. in the viable limits, the variation even in heavy exercise is only 45 per cent.

A very interesting point in the estimations of Dill, Talbott and Edwards is the range of normals. They lie between $pH = 7.36$ ($cH = 4.4 \times 10^{-8}$) and $pH = 7.44$ ($cH = 3.6 \times 10^{-8}$), a variation of range of only 22 per cent. This range is scarcely greater than that of the maximal daily variation in a single individual as found by Cullen and Earle (1929). The blood may become more alkaline towards evening to the extent of seven-hundredths of a pH.

COMPARISON OF THE CONSTANCY IN MAN WITH THAT IN LOWER ANIMALS

With these figures let us contrast those from the fish. Determinations, kindly given me by Dr Hall and Dr Gray of Duke University from the blood of the scup, taken at rest, showed a variation in the hydrogen-ion concentration of the blood between cH of 3×10^{-8} and of 2.5×10^{-7}, *i.e.* $100 : 833$, a variation not of 22 per cent. but of 800 per cent. The extreme limits of hydrogen-ion concentration in the blood of sub-mammalian forms are given differently by different authors and are quite obscure. Rohde (1920) believes that frogs normally have a pH which varies from 6.32 to 7.13, and that if fed on boric acid it may fall within 10 minutes to 4.2, while if fed on soda it will rise to 8, *i.e.* a ten-thousandfold variation. These figures, however, have not been substantiated by Mrs Hertwig-Hondru (1927), who placed the variation within much narrower limits (7.36 to 7.59). Even these are much larger variations than for man when at rest normally.

The following figures (Table II) are given for the hydrogen-ion concentration in the insects cited (Glaser, 1925):

Table II

	pH	
Grasshopper	7·2	7·6
House-fly	7·2	7·6
Cockroach	7·5	8·0
Malacosoma	6·4	7·4
Bombyx mori	6·4	7·2
General range	6·4	8·0

The evolution of the whole mechanism which participates in the regulation of the hydrogen-ion concentration of the blood is a matter of extraordinary interest. It concerns primarily the kidney, the blood itself, and the respiratory centre.

The blood. The evolution of the blood has been such that in general the more highly developed the form of life the less does the addition of a given quantity of acid or alkali alter the hydrogen-ion concentration, that is to say the more perfectly is the blood buffered. This is only true in a very rough sense. For this purpose the animal kingdom may be divided into great blocks: (1) the sub-vertebrate forms, (2) the cold-blooded vertebrates, and (3) the warm-blooded vertebrates, the birds and mammals.

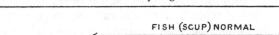

Concentration of hydrogen ions in blood

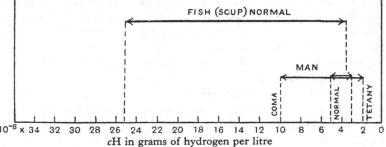

Fig. 2

Moreover, there are three principal ways in which the addition of acid to the circulating fluid is prevented from suddenly and unduly altering the hydrogen-ion concentration of the same:

(1) By the buffering of the fluid and the tissues in contact therewith.

(2) By the excretion of acid or retention of alkali by the kidney (and the production of ammonia).

(3) By the excretion of carbonic acid by the lung.

It will at once strike the reader that the first method is not quite "on all fours" with the second and third; the first is rather a method for the mitigation or "evasion" of the effects of the addition of acid and alkali, than for the regulation of the number of hydrogen ions present. This distinction is one which I will refer to; here it is not of great importance, but please note it for future reference.

In the circulating fluid in such situations as the water vascular system of the sea urchin there can be very little buffering (see Fig. 3). In many of the sub-mammalian forms the blood is buffered to quite an appreciable extent. This buffering is associated with the acquisition of some sort of pigment for the purpose of carrying oxygen, the two chief of which are haemoglobin and haemocyanin. Haemoglobin is of course found in considerable quantities in the blood of many worms, while haemocyanin is the prevalent respiratory pigment in the arthropods and molluscs.

It is not to be supposed that these pigments had originally any function in the blood other than that of carrying oxygen, if indeed they had any respiratory purpose whatever. But in each case it was desirable, in order that the pigment should be as useful as possible, that the oxygen should be capable of the most easy detachment in the situations in which it was most badly needed. Such situations would in the main be those in which carbonic acid or some other acid was likely to accumulate, and therefore it has come to pass that these two pigments have survived. The buffering action appears to be purely incidental, and indeed it is doubtful whether in such low forms of life there is any particular object in having the blood very highly buffered.

So far then as haemoglobin is concerned (and the evolution of the principal buffer is the evolution of haemoglobin), the consideration of buffering appears to have been somewhat of an afterthought, for recently Redfield and Florkin (1931) have discovered in a worm, *Urechis caupo*, a form of haemoglobin in which carbonic acid has no effect on the affinity of the pigment for oxygen.

Similarly, the most primitive form of haemocyanin, like its counterpart in the haemoglobin series, does not appear to have any value as a buffer. The oxygen affinity of the more complicated forms is affected by carbonic and other acids.

With the transition from the lower to the vertebrate forms came the next and indeed the final stage in the evolution of a highly buffered internal medium for the body. That stage was the enclosure of the haemoglobin in corpuscles. It is not that corpuscles

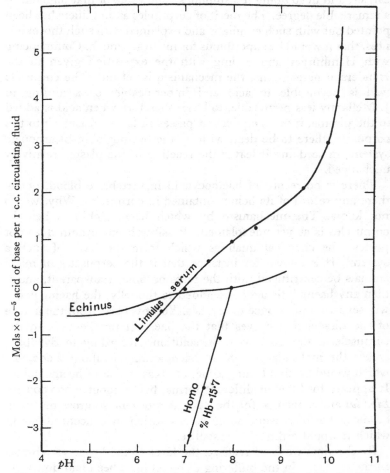

Fig. 3. Titration curves of blood of Man (after Terroux), of *Limulus* (after Redfield) and of *Echinus* (after Pantin).

which contain haemoglobin are unknown in lower forms of life, but putting on one side the exact phylogenetic status of *Amphioxus*

—the vertebrate does, as the invertebrate did not, systematically and extensively use intracorpuscular haemoglobin as its oxygen carrier, and in so doing it "amplifies" the value of that pigment to a remarkable degree. The merit of corpuscles as amplifiers has been pointed out with such emphasis and explained with such thoroughness, that it would be superfluous for me to say much. Commencing with Hamburger and ending with the exposition given in the Henderson nomograms, the mechanism is set out. The corpuscle wall is permeable to acid and impermeable, or according to J. Mellanby less permeable, to base; therefore when acid is added to the plasma, it or its equivalent passes in large measure into the corpuscle, there to be dealt with by the haemoglobin-bicarbonate system; in so doing it leaves the reaction of the plasma relatively unchanged.

There is no record of haemocyanin in vertebrate blood, nor is there any record of its being contained in corpuscles. Why, we do not know. The mechanism by which haemoglobin is held in corpuscles is as yet unexplained. Possibly haemocyanin does not possess the chemical qualities which form the basis of such a system. It is known for instance that if the corpuscles of many animals be centrifuged until the mass becomes transparent, and if then any haemolytic agent be added to the jelly, the haemoglobin will separate out at once in crystals. Yet it is clear from the nature of the dissociation curves that the haemoglobin behaves in the corpuscle as though it were in solution. According to Svedberg (1926) the molecular weight of haemocyanin is about 2,000,000, which would be about thirty times as great as that of haemoglobin. It appears to differ in different forms, being about 2,000,000 for *Limulus* and 5,000,000 for the snail. A molecule so gross may well be incapable of masquerading as a solution in concentrations in which it would ordinarily crystallise.

By the time we have reached the level of the lower vertebrates all the elements in the buffering of blood have been laid down. As regards buffering there is little difference in kind between frog's blood and that of an anaemic human being. The transition from the cold-blooded to the warm-blooded vertebrate is a matter of degree.

The preliminary step is that of producing an internal medium into which acid and alkali can be put with but a disproportionately small alteration in reaction; the finer points as regards the regulation of the hydrogen-ion concentration depend not upon the blood itself but upon the organs of excretion. As the excretion involves both substances in aqueous solution and carbonic acid gas the organs in question are the kidneys and the lungs respectively.

The kidney. The primitive organ for the regulation of hydrogen-ion concentration as of other things in the blood is the kidney.

I suppose the great problem which the kidney presents is: How can it regulate the blood concentration of so many substances simultaneously? Whatever the answer to this question may be, it would *a priori* not be surprising if in the effort to do a great many things at the same time, the kidney did each one of them a little less perfectly than would otherwise be the case. That may be the reason why, if the composition of the blood is altered by the injection of acid or alkali, a long time elapses before the previous *status quo* is entirely re-established.

Suppose there are 4·8 litres of blood in the body, and that the volume of blood which traverses the kidney is 2 c.c. per gm. per min., and the weight of the kidneys 320 gm., 640 c.c. of blood per minute will pass through these organs, or 13 per cent. of the blood in the body. Suppose (1) that a gram of some foreign substance, such as bicarbonate, is injected into the blood and only escapes therefrom in the kidney, and (2) that the blood leaving the kidney is entirely freed from this material, in 5 min. the amount in the blood will be halved, and in 20 min. there will only be about one-tenth of a gram remaining.

In point of fact the above assumptions do not hold. In general, material which can be eliminated by the kidney can also pass into the tissues, to be returned later to blood flow.

We may, however, "bracket" by making an assumption equally extravagant, namely that the alkali immediately diffuses all over the body, so that the whole body gets into equilibrium before any alkali is secreted; but let us retain the assumption that the kidney is as efficient as possible and that the blood emerged in the renal vein normal. On the above assumption, as the flow through the kidney per minute represents about 1 per cent. of the body weight (the water in that blood will be rather more than 1 per cent. of the water in the body) the excess of bicarbonate should be reduced to half in about an hour and to 10 per cent. of its amount in something of the order of 4 hours. One might expect then that the actual time for the elimination of alkali would be somewhere between that given by the assumption that all the alkali stayed in the blood and that all diffuses at once into the tissues and that in a time which was not less than 20 min. nor more than 4 hours the excess of alkali would be reduced to one-tenth of its original amount.

But in point of fact nothing of the kind occurs. Davies, Haldane and Kennaway (1920) showed that when a considerable amount of alkali (57½ gm. of sodium carbonate, within about 20 min.) was ingested, after 9 hours the kidney was still eliminating sodium carbonate in large quantities; after about 12 hours the elimination ceased.

It would seem that our second assumption must go too, as the kidney does not reduce the venous blood to normality and is therefore not as efficient as it might conceivably be.

The kidney apparently works—so far as alkali is concerned—on quite different principles. The authors quoted made the illuminating observation that at best the kidney produces (presumably because it can do no more) urine of a certain limiting concentration of alkali (about 2·5–2·8 per cent.). Any further increase in the elimination of alkali beyond that obtained by raising the content of the normal volume to 2·8 per cent. must be effected by increasing the elimination of water. In that case either the concentration or the amount of every soluble substance in the body becomes affected indirectly. Short of knowing that the kidney cannot eliminate urine of more than a certain alkaline content we are very much in the dark as to any quantitative statement regarding the factors which regulate the elimination of alkali by the kidney. There seems, however, to be good reason for supposing that the influence of the nervous system does not figure prominently as one of them.

Many researches have of course been made upon the effect of the nervous system upon renal secretion, but none of them gives information as to the rate at which a given excess of alkali or acid is rectified when the kidney is denervated. Margaria (1931), there-fore, undertook such experiments. The general plan was that the kidney on one side was denervated by a preliminary operation; after a suitable time (10-45 days) the animal was rendered insensible either by chloralose, or in the later experiments by decerebration, cannulae were inserted into the ureters close to the bladder, and after sufficient measurements of the quantity and acidity of the urine from each kidney had been made, alkali was given, or in some cases carbonic acid. Taking the period before the actual administra-tion of alkali, Margaria found, as others had done, that the excre-tion of water was more rapid on the denervated side; he found also that the excretion of acid was more rapid, too, but not in proportion to the rate of excretion of water. These effects were most marked in the experiments in which the kidney had been most recently denervated. In those in which a month or more had elapsed be-tween the denervation and the final observations there was little difference between the two kidneys.

What has been said about the kidney has assumed that the function of that organ in relation to the regulation of hydrogen-ion concentration is one of selection; that it withdraws certain in-gredients from the blood and at the same time eliminates them. No doubt in the main the picture is correct. Nevertheless, the work, especially of Nash and Benedict (1921), has led to the belief that the kidney manufactures one very strong base, ammonia, out of a neutral substance, urea. Not only so but that it can throw the ammonia so made or some of it into the circulating fluid, so that the blood in the renal vein is richer in ammonia than that in the renal artery. There appears here to be a most interesting field for the comparative physiologist.

The lung. The activity of pulmonary respiration is of course conditioned by that of the respiratory centre. Our conception of the evolution and mechanism of the respiratory centre must at the present time be considered in the light of the work of Lumsden. His position may therefore be shortly stated. Lumsden (1923) postulates at least four centres situated at different levels in the brain. Of these the lowest is near the calamus scriptorius. It occupies the position of the classical *noeud vital.* It is also phylogenetically the oldest. If I have followed Lumsden aright, he regards it as only a survival in vertebrates even as low as the tortoise. It is a gasping centre, and presumably even in the tortoise it must be regarded as an emergency mechanism. The interesting point about the gasping centre is its apparent unresponsiveness to CO_2. It appears solely to be actuated by want of oxygen. The animal takes in a gulp of air, and when the oxygen in this is exhausted it takes in another. In this connection, recollect that the main exchange of CO_2 in the frog is through the skin (Krogh, 1904), the function of the lung being primarily concerned with the acquisition of oxygen. Therefore the connection of the respiratory centre with regulation of a constant hydrogen-ion concentration is an afterthought; as in the case of haemoglobin the sequence of function is: oxygen intake, CO_2 output, regulation of constancy of hydrogen ions.

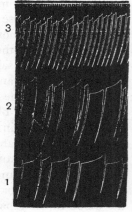

The next step in the evolution of the respiratory centre is typified by the brain of the tortoise. In it we have a response both to oxygen want and CO_2 excess, but one which admits of rather considerable alterations in reaction. Want of oxygen increases the frequency of the rhythm, excessive CO_2 (30 per cent.) principally affects the rate, increasing the force of expiration. According to Lumsden (1924) respiration is operated in the Chelonian brain by three centres, each of which is situated in the medulla at the level of the fourth ventricle. Of these centres the principal is that which he calls the apneustic centre. Its activity consists in inflating the lung with air and keeping it so inflated. But (still following Lumsden's description) when the hydrogen-ion concentration in the blood (seen from Gesell's angle in the centre itself) reaches a certain limit, the inspiratory effort is broken down and a passive expiration takes place. This passive expiration over, the apneustic centre gets once more

Fig. 4. Apneuses in the tortoise (after Lumsden). Upwards = inspiration, downwards = expiration. Reads from right to left. 1, normal; 2, breathing nitrogen; 3, breathing 30 % CO_2.

to work and another inspiration takes place. The second centre is an expiratory centre which comes into action simultaneously with the breakdown of apneusis and from the same cause. The centre augments the passive expiration, and that to a greater or less extent according to circumstances. Thirdly, there is the gasping centre which according to Lumsden is not ordinarily used.

As Lumsden points out, the arrangement which serves for the tortoise would be quite inadequate for the higher mammals. Among other things, each improvement in the buffering of the blood would tend to reduce its power to break down apneusis, and therefore the organism would tend towards a chronic condition of oxygen want. An additional mechanism has been therefore evolved which is found in the dog, cat and rabbit and presumably in higher mammals; a fresh centre has been added still higher up at the level of the upper part of the pons. Apneusis is now broken down by the action of the vagus on and through this higher "pneumotaxic" centre. As the stimulus to the vagus is regarded as the passage of air through the lung, caused by the very fact of apneusis, a type of respiration is effected in which the two phases follow immediately the one upon the other. Increased hydrogen-ion concentration in the blood continues to have a rôle, which is not to initiate one or another phase of the rhythm, but to regulate it—especially in the matter of depth.

The above description of the evolution of respiration is, as nearly as I can set forth, that of Lumsden. Let me now make some comments upon it.

First of all I wish to separate the actual phenomena which he describes from the rigid reference of these phenomena to precise anatomical centres. It seems to me that when a section affects a particular phase of respiration the deduction is no more than that some portion of the complex paths involved has been severed, and it is possible that if the animal were kept alive an alternative path might take its place. This criticism is put forward with great reserve from one who is not a neurologist.

Putting anatomical questions right out of the picture there remains the question in which I was much interested, whether Lumsden's physiological levels can be reproduced in ways other than sections. Of such methods I will mention two: the first is the respiration of hydrocyanic acid gas, the second is the study of foetal respiration.

A considerable research on the effect of HCN on the respiratory centre has recently been carried out by Taylor (1930, 1931). By the administration of air containing small quantities of the gas, the effects develop so slowly as to take 30 min., or a time of that order, to produce death. Moreover, they can be interrupted and reversed at any point. Lumsden's typical phenomena are the normal respiration, the apneusis and the gasp—in that order: at the commencement of poisoning the normal or pneumotaxic respiration turns into a series of apneuses, and these later turn into gasps (Fig. 5). Any discussion of the transitional forms I defer for a moment.

The deduction from Taylor's experiments is that the gasp is the fundamental phenomenon of respiration. It appears to be unaffected by carbonic acid and therefore Taylor agrees with Lumsden that the fundamental movement proper to pulmonary respiration is not one which is prompted either by carbonic acid or hydrogen ions. Lumsden's observations evidently have a physiological significance and are not merely anatomical artifacts.

At this point let me introduce the research of Huggett (1930). Huggett studied the physiological processes in the embryo. In foetal goats 2 months before their full term there are no respiratory movements; at least he was quite unable to confirm the work of Ahlfeld (1890) who, in opposition to Zuntz, held that such occurred. Starting then with the condition of quiescence, it was possible to induce single inspirations by one or other of two methods: (1) by stimulation of the central end of the sciatic nerve, and (2), as had previously been shown by Pflüger (1868) and Zuntz (1877), by clamping the umbilical cord. Cutting the vagus at this stage had no effect, as indeed might be expected in a condition in which the respiration was not of the pneumotaxic type. That stimulation of a sensory nerve such as the sciatic could institute respiratory movements in a quiescent respiratory centre is, however, a matter of great interest.

The nature of the gasp. Approaching this subject by three different avenues, those of brain section, cyanide poisoning and foetal development, we arrive at the "gasp" as being the fundamental phenomenon of respiration, and by two avenues, at the apneusis as being a sort of half-way house between ordered respiration and the

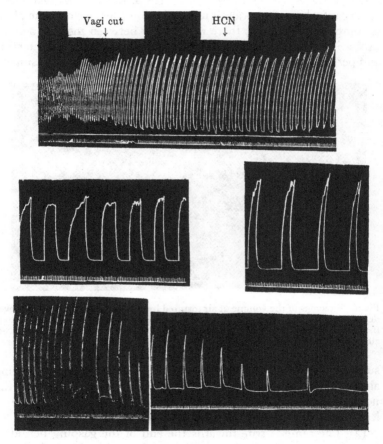

Fig. 5. Respiration of cat during continuous inhalation of air which contains HCN. (H. Taylor.) Reads from above downwards. Uppermost tracing shows pneumotaxis; middle row, apneusis; lower row, gasping.

gasp. It therefore seems worth while to enquire more particularly what may be the true nature of the gasp and of the apneusis. In the first place a gasp is a unit, it seems unlikely "to have any previous history in the central nervous system", if I may use a phrase borrowed from Sir Charles Sherrington. In support of this contention three reasons may be given.

(1) In the marmot, during winter sleep, respirations take place at very infrequent intervals. The respiration when it does take place is complete of its kind, there is no delay between inspiration and expiration, but the inspirations occur at quite irregular intervals and perhaps 4 or 5 min. apart (Fig. 6). It does not seem feasible to

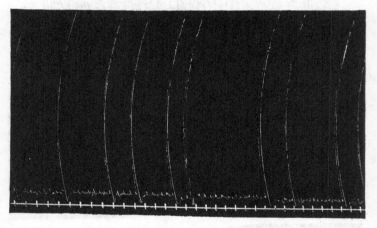

Fig. 6. Tracings of the volume of air inspired by the marmot. A rise of the lever denotes inspiration. The marmot was in an air-tight box, fitted with a spirometer, the movements of the lever of which were recorded. It breathed (through valves) air from outside the box. Time, ½ minutes. (By permission of the Royal Society.)

suppose that after the end of one respiration a reflex is drifting about the central nervous system for 5 min. before it prompts the next. Even if it did so the respirations might be expected more or less regularly at 5 min. intervals (Endres and Taylor, 1930).

(2) The same is true towards the end of the gasping period of cyanide poisoning. Each gasp is complete of its kind, fairly reproduces its predecessor and the expiratory phase follows the inspiratory phase with precision and without delay, but the gasps take place at rare intervals and irregularly (Fig. 7). Obviously the causal relation between an inspiration and the subsequent deflation of the

chest is quite different from, and of a much more intimate character than, the causal relation between expiration and the commencement of the succeeding gasp.

(3) A foetal goat of 10 weeks (the period of gestation being 21 weeks) I have seen to make as perfect a gasp as that of a cat dying under cyanide. The foetus was attached to the mother by the cord,

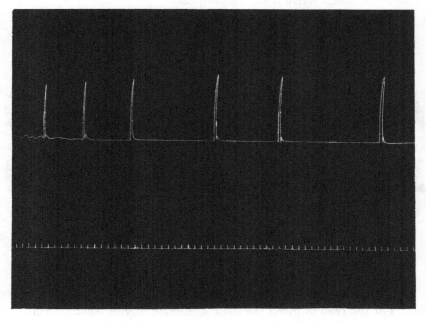

Fig. 7. Gasps in cat towards the end of cyanide poisoning.

but was lying in a bath of saline, having been withdrawn through a Caesarian section. The gasp was an isolated one and was complete in all its phases.

So much for the gasp as a unit. With regard to the fusion of gasps into a rhythm: physiological rhythms, as A. V. Hill (1933) has pointed out, are of the following general nature. A continuous process, *e.g.* the filling of a cistern, proceeds until some crisis takes place, *e.g.* the syphoning of the contents at a rate more rapid than the continuous flow. The cistern empties and does so at regular intervals.

The beautiful researches of Adrian and Buytendijk (1931) seem to indicate that in the goldfish and even in much lower forms of life respiration is a manifestation of an inherent rhythm in the central nervous system, and does not depend upon afferent influences which arrive from the periphery.

From the central nervous system, even after complete removal from the body, a rhythm may be tapped electrically and registered, which corresponds with that of the rate of movements of the goldfish gills (Fig. 8).

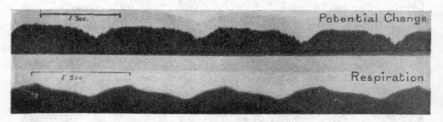

Fig. 8. Potential changes in excised central nervous system of goldfish compared with respiratory rhythm.

The gasp, apart from the property already discussed, namely that it is a separate entity, appears to have two others which may be noted here. It is maximal, involving probably all the respiratory muscles possibly in full degree, and it is very rapid.

For the reasons given the most legitimate way in which to consider the gasp (or at least its *nervous* mechanism) seems to me to be as a unit, which begins at the commencement of inspiration and which at the end of expiration is finished with. And now to study the gasp in greater detail, we may turn to the researches of Adrian, of H. Taylor and N. B. Taylor (1931). These researches consist in a study of the contractions of the muscles of expiration by the method of tapping the electrical variations which take place when they contract; the electrical variations are amplified and turned into sound waves. They can then be heard, so that by listening to the sounds emitted from a "loud speaker" an opinion can be formed as to whether in any particular phase of respiration the muscles of expiration are called into action. When this technique is applied to the triangularis sterni during gradual cyanide poisoning, we learn that as pneumotaxis passes into apneusis the triangularis sterni ceases to contract during expiration, and that during the periods both of apneusis and of gasping, costal expiration as an active process does not, or does not necessarily, occur (Fig. 9). The deflation of the chest is purely passive. The abdominal muscles tell the same

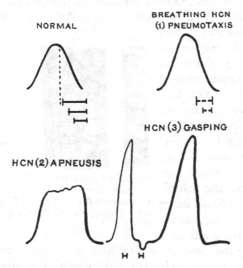

Fig. 9. Schematic representation of sounds heard in triangularis sterni muscle during expiration. ⊢⊣ =loud sound. ⊢⋅⋅⊣ =feeble sound.

tale. We arrive at the conclusion, therefore, that a gasp is an inspiratory effort which, untrammelled, passes over the whole, or nearly the whole, of the nervous centres actuating the muscles of inspiration—both those of normal inspiration and those of forced inspiration. Having traversed them rapidly "and without let or hindrance" the wave is played out, the muscles concerned cease to contract and the chest becomes deflated.

Relation between the apneusis and the gasp. I have said, "without let or hindrance", and the phrase stands for more than a mere rhetorical embellishment. There appear to be two distinct physiological processes involved in normal expiration, firstly the checking of inspiration and secondly the contraction of the expiratory muscles. Very often even in normal respiration they do not commence synchronously. In order of time the summit of the curve of respiration is signalised by the complete checking of inspiration (*i.e.* the relaxation of the inspiratory muscles), and the expiratory phase may be half over before the triangularis sterni or any other muscle which we have tapped commences to contract. Both these processes are abolished during the gasp, and the wave of inspiration goes "the whole way". Not only in a single respiration do the checking of inspiration and the appearance of active expiration not take place synchronously, but the latter process is abolished at an earlier stage in cyanide poisoning. In apneusis, the expiratory reflex has already been abolished, the inhibitory reflex is gradually being weakened. If a normal respiration be regarded as a

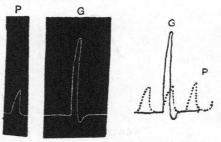

Fig. 10. Showing the relation (1) in amplitude, (2) in rapidity between the ordinary respiratory movements P and the gasp G, taken from the same tracing of cat inspiring air containing HCN.

modified gasp the process of inhibition must commence almost as soon as the inspiration itself. It is only necessary to superpose, as in Fig. 11, (1) a normal respiration, (2) an apneusis and (3) a gasp, taken from an animal which is "breathing" cyanide to see that the curve rises most slowly in the pneumotaxis and most rapidly in the gasp. Indeed on the tracing of apneusis actual breaks can be seen which correspond to checks in rate in the development of inspiration. In the case of apneusis, the inspiration is apparently inhibited at a certain point and a struggle takes place for a time between the progressing wave of inspiration and the presumably reflex inhibition which the wave brings into being; if the two are nicely balanced the apneusis lasts for a considerable period before the wave is spent and the chest deflated. It will be seen that, while we have confirmed Lumsden's phenomena, our interpretation of them differs from his in one important respect. He believes the gasping centre to be entirely superseded by the apneustic centre, we visualise an apneusis as being a modified gasp. That difference of opinion is based on information not at Lumsden's disposal, namely the gradual transition from apneusis to gasping when cyanide is administered, or from gasping to apneusis when cyanide is withdrawn. Fig. 12 shows such a transition, in which the last short apneuses alternate with gasps. It does not seem possible to regard the two as being due to the alternate workings of two different and independent centres. Such an idea would, I think, be very far fetched.

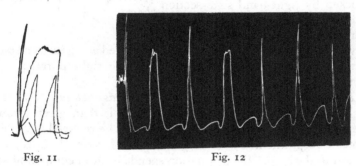

Fig. 11   Fig. 12

Fig. 11. Normal respiration, apneusis and gasp superposed.
Fig. 12. Apneuses passing into gasps. HCN. (Taylor.)

The simple interpretation is that in the case of the apneusis the gasp is interrupted. The interruption is not so complete as immediately to break down the inspiration; a struggle therefore ensues for a short time in which the inspiration spends itself in competing with the inhibition.

I am indebted to Adrian for pointing out to me that though at first sight it might appear unlikely that the inhibition should do other than completely abolish the positive phase of the inspiration, there is perhaps an interesting analogy in the work of Bethe (1930) and Bethe and Woitas (1930), who give numerous instances from beetles and other forms in which a simple purposive movement thwarted along the direct and usual line of attainment is not stopped, but persists to achieve its end by some unusual path. Indeed, Sherrington has pointed out to me that it is unnecessary to go so far afield. I quote his words: "An intriguing example of this (a simple purposive movement when thwarted along its direct and usual line of attainment taking resort to an unusual one) with regard to respiratory ventilation itself is W. T. Porter's (1895) phrenic experiment: discharge by the right phrenic is blocked by right-hand semi-section (above the 3rd cervical) until the left phrenic is cut". The impulse coming down the left side of the cord, being thwarted along its usual path, the left phrenic, then pushes across to the right phrenic along which it discharges.

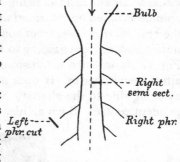

The relation of pneumotaxis to the apneusis and the gasp. The logical inference from the above argument is that, given a much greater degree of inhibition, the gasp would be reduced to a pneumotaxic respiration. So far so good, but I have said above, with regard to the gasp, "the inspiration spends itself in competing with the inhibition". But if the inhibition is increased to the point of suppressing the inspiration, the questions may reasonably be asked, Will the inspiration spend itself completely, and if not what will happen when the inhibition is removed? Perhaps I can picture the matter in the following way: suppose a gasp to result from 100 explosions in each of 100 cells in this centre, after which the cells rest till the system is re-established, and supposing in pneumotaxis only 30 explosions per cell take place and fewer cells are involved, say 50, then 1500 explosions only will have occurred and most of the explosive material will be left ready to go off as soon as the inhibition is removed. Will they commence spontaneously to do so? If so, a new and much more rapid rhythm will at once be established, and that actually occurs.

In the light of more recent additions to knowledge we arrive at a conception of respiration which is a little different from that of

Lumsden. A respiration commences with a wave of activity in the centre responsible for inspiration. This inspiration has all the potentialities of a gasp, but at once induces an effort to smother itself and it is broken down. Firstly there is inhibition of inspiration and secondly active expiration. Cyanide abolishes both these, but it abolishes the latter before the former. When both are gone respiration is reduced to its simplest and most primitive terms, namely a series of gasps. When the inhibition is reduced to very feeble proportions the respiration conforms more or less to the apneustic type.

The action of carbonic acid. Now to turn to the action of carbonic acid. As we have already said, CO_2 appears to have no augmentor effect at the gasping stage. How could it? Three possible ways suggest themselves:

(1) It might increase the inspiratory efforts, but it cannot effect anything in that direction because they are more or less maximal in any case.

(2) Were expiration an active process CO_2 might affect it, but in the case of the gasp, expiration being purely negative, there is nothing to affect.

(3) Though it cannot exert much influence on the nature of the gasp it might still reduce the time which elapses between the gasps. This it appears not to do. Here then is one perfectly definite piece of information, which though negative in character is well worth noting, because we can trace the characteristic of CO_2 right up to man. Short of an increase of CO_2 of the order of 5 per cent. in the inspired air it does not necessarily quicken respiration. In some persons CO_2 quickens respiration slightly, in some it has the opposite effect, in yet others it has no effect at all.

Haldane and Priestley (1905) found this in their classic investigation on the effect of CO_2 on the respiratory centre, and it has been the experience so far, I know, of others who have interested themselves in the matter. When we speak of carbonic acid stimulating the respiratory centre we mean or should mean that CO_2 (within ordinary limits) increases the amplitude immediately and the frequency ultimately. This is perhaps worth a word of emphasis because one constantly hears persons discuss the effect of CO_2 as being quickening of the respiratory rhythm (see Barcroft and Margaria, 1931). The introduction of the factors which cut "gasping" down to pneumotaxis provides the possibility of a new milieu in which CO_2 can work. Fig. 13 shows that in the gasping phase of cyanide poisoning the total ventilation is much greater than in the phase of pneumotaxis.

In the hibernating marmot the effect of CO_2 seems to be rather simple (Endres and Taylor, 1930).

(1) In the first place it postpones the crisis which turns inspiration into expiration. The curve B in Fig. 14 (marmot breathing

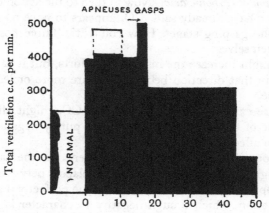

Fig. 13. Minutes during which HCN 0·31 g./m.³ is administered.

air + 10 per cent. CO_2) rises no more steeply than curve A (marmot breathing air), but it proceeds further before it is checked.

(2) The expiratory phase, especially the latter part of it, is much accentuated. The respiration is complete before the normal.

CO_2 acts quite differently from rise of temperature, cf. C and D.

In man the effect of CO_2 is not so simple. Fig. 15 shows tracings taken by rebreathing into a Krogh's spirometer; the modifications which result are due to the breathing of the carbonic acid in the concentrations stated in the legend.

In man the curve of inspiration is much steepened by carbonic acid. That new factor, which provides some justification for the statement that "CO_2 stimulates the respiratory centre", demands a closer enquiry into the effect of CO_2 on the muscular contractions involved in respiration.

The effect of CO_2 on the diaphragm. In the central part of the diaphragm during normal respiration there is a condition of tone throughout the whole of respiration, heightened of course during

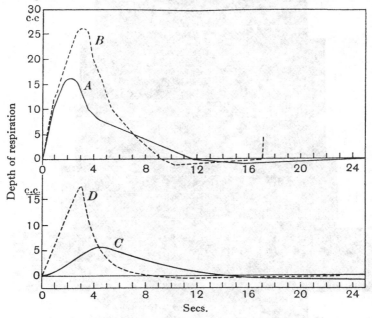

Fig. 14. A tracing of the volume of air breathed by hibernating marmots. The animal used for *A* and *B* was not the same as that used for *C* and *D*. Tracing *C* reaches the base-line at 45 seconds from the commencement. Temperature of marmot in *C*, 4·8°; in *D*, 19·6°. (By permission of the Royal Society.)

inspiration and reduced during expiration. It can easily be appreciated by the Adrian technique; indeed I possess a gramophone record, made under Adrian's instructions by the Columbia Company, in which the muscular contractions in some parts of the diaphragm can be heard during the whole of normal respiration. The sounds wax during inspiration and wane during expiration.

When the animal "rebreathes", the sounds become very much

accentuated during inspiration and abolished during expiration. Presumably, therefore, the amplitude of the movements of the diaphragm increase in both directions*.

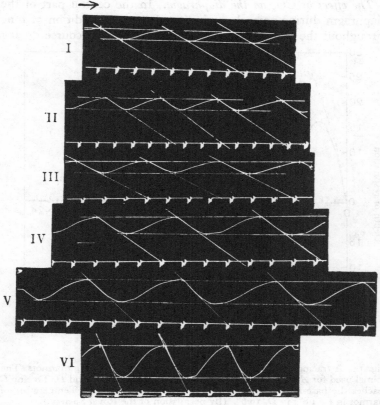

Fig. 15. Spirometer tracings of the respiration. Subject, Barcroft. CO_2 in inspired air: (i) 0·2 per cent., (ii) 1·0 per cent., (iii) 2·2 per cent., (iv) 4·2 per cent., (v) 5·3 per cent., (vi) 7·5 per cent. Inspiration downwards, expiration upwards. Tracing read from left to right. Time, 1 sec.

The effort to visualise what takes place during diaphragmatic respiration is perhaps allowable, even with no more information than we possess.

* This record was "exhibited" at the Dunham Lecture in Boston in October 1929.

Let us start with a diaphragm in a state of partial tone; connected with it is a cell R, Fig. 16 (one of a great number of such), in the gasping centre. (I omit connections at the level of the phrenic nucleus.) Were there no more machinery the respiration would be of the gasping type, regular in rhythm, the frequency being that

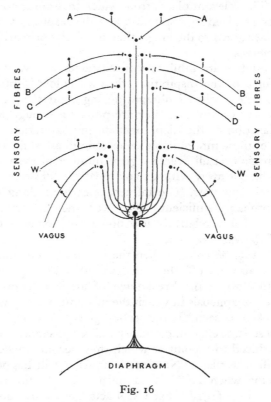

Fig. 16

of the cell R. The total ventilation would be much greater than that of normal respiration. The gasping centre left to itself would produce hopeless over-ventilation. Too many cells, in fact probably all the cells (for gasps seem to be maximal), would "fire off" at the same time or in quick succession. The gasp must be checked, and this inhibition is excited from above by some mechanism the

activity of which is (1) impaired by section of the so-called pneumotaxic centre, (2) destroyed by HCN, (3) relatively undeveloped
in the foetus, (4) heightened by the stimulation of sensory nerves.

Inspiration once it is initiated evokes 10,000 (figuratively, probably more actually) sensory impulses which immediately tend to
throttle it. The relevant places from which they come are probably
the muscles involved in respiration and the respiratory tract from
the nose downwards to the lungs, thus involving primarily the fifth
and tenth nerves.

These impulses are inhibitory in character and vary in degree
with the force of the inspiration, thus a shallow inspiration will
reduce the tone of the diaphragm during expiration, a deep one
will abolish it. In any case they are powerful enough to account
for the relaxation of the diaphragm during expiration. Probably
only a few of these innumerable reflex stimuli are sufficient to preserve respiration, at all events a great number of avenues of sensation can be cut away without any great modification of the respiratory ebb and flow. The influences coming along the vagus
alone are probably sufficient. Moreover, the connections of the
vagus are probably relatively low down, for apneusis cannot be
obtained if the vagi are intact.

But if the vagi be cut and then the impulses from above points
of entry are also cut off, the remaining afferent impulses are insufficient to enforce the breakdown of an inspiration, and the
phenomenon of apneusis in which the inspiratory effort is checked,
but not broken, is seen. If the section goes below the "apneustic
centre" these last remaining sensory reflexes disappear, or at least
they get reduced to a point at which they become ineffective and
gasping will take place. Now consider what will happen on the
above scheme when HCN is administered. On the theory put
forward by Evans (1919), that HCN acts from above downwards,
cells above R will be the first to be affected. As the finger is gradually
removed from the throttle so the activities of R will become less
and less restrained, the amplitude of the respiration will increase
and hyperventilation will, as indeed it does, take place. This is the
first phase of HCN poisoning, the so-called "stimulation" of
respiration. But from our standpoint it is not a stimulation in any

true sense, it is the removal of an inhibition. It is very difficult to think of HCN really as "stimulating" anything: it is a general protoplasmic poison. It is unnecessary to detail the other phenomena, they would happen in time just as Taylor found them to do.

Costal respiration. The essential difference between diaphragmatic and costal respiration lies in the fact that the diaphragmatic muscles are purely inspiratory. The costal muscles are both inspiratory and expiratory. Concerning expiration we know but little. That there must be an expiratory centre is, I imagine, agreed by everyone now. That the expiratory centre is thrown into activity as the result of inspiration is also, I think, clear. In HCN poisoning the electric variations of the expiratory muscles are occasionally very well marked just after the gasp is over, so that the impression is given of an inspiratory gasp followed by an expiratory one. How far it is possible to apply to the expiratory centre the same arguments with regard to inhibition that we have to the inspiratory centre it is difficult to say. We can easily get inspiration without active expiration, we do not get active expiration without inspiration—in mammals at all events.

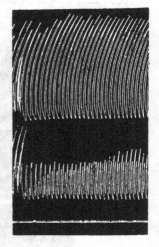

Fig. 17. Double abdominal movements (after pumping air) (cat). Top: thorax. Bottom: abdomen. Vagi cut. Time in sec. HCN = 0·28 gm. per cu. metre. The interpolated movements on the abdominal curve are caused by active expirations, the abdomen being forced outwards by the contractions of the thorax.

If the hypothesis that CO_2 acted like making suitable cuts through the medulla or like the administration of HCN were true, there seemed to be just a chance that the giving of enough CO_2 might produce the chain of symptoms, hyperpnoea, apneusis and gasping, which is characteristic of the other procedures. The event proved more successful than we had expected. Naturally if you set out to imitate HCN with CO_2 there is no use adopting half measures. So we gave the cat a mixture of 64 per cent. CO_2 and

28 per cent. oxygen, with the result that hyperpnoea at once supervened, to be followed by an obvious tendency to respiration of an apneustic type, this in turn passed into "gasping" (Fig. 18). There are differences in detail from HCN and also from the types as given by Lumsden, but it must be remembered that the action of HCN and CO_2 is at best differential. Their action on the higher portions of the brain is not complete before that on the lower

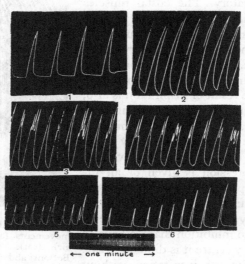

Fig. 18. Effect of CO_2 64 per cent. on respiration of cat.

1, Normal   3, Apneusis   5, Gasping (imperfect)
2, Hyperpnoea   4, Apneusis (later)   6, Gasping

parts commences. Indeed an attempt to obtain the whole chain of events produced (*a*) from sections, and (*b*) from HCN, by the administration of CO_2, proved to be much more successful.

The important phenomena of diaphragmatic respiration have, I think, been accounted for with one exception, namely, that the rhythm is considerably faster than the natural gasping rhythm, which presumably is that proper to cell *R*. To that I will revert after saying a word about costal respiration. Accepting the general scheme of respiration as put forward above there is an alternative explanation of the action of CO_2 as was pointed out to me by Dr Margaria. It is as follows: carbonic acid stimulates at the level of R. During gasping the stimulus effects nothing because the gasps are maximal, but when R is subject to the normal inhibitions which play upon it, CO_2 causes it to "put up" a stronger fight against these inhibitions. This conception introduces the extra complication of taking CO_2 out of line with other things (HCN, etc.). Moreover, there is something on the positive side to be said for the idea that the effect of CO_2 is rather to remove inhibitions than to cause active stimulation.

According to Heymans (1931) the respiration is easily affected by influences which reach the brain from the carotid sinus. When the carotid sinus of a dog B, and that only, be perfused with blood from a dog A, the respiration of B will be increased in depth and rate:

(1) if the blood pressure of dog A be lowered;

(2) if dog A breathes HCN;

(3) if dog A breathes CO_2;

whilst respiration in dog B is made shallower and slower:

(4) if the blood pressure in dog A be raised;

(5) if CO_2 be removed from the blood of dog A.

As between (1) and (4), the raising of the blood pressure is more likely to be a stimulus than the lowering of blood pressure—a negative act; and with regard to (3) it is much easier to conceive of CO_2 as paralysing a nervous process than as stimulating it. The discharge in the cardiac depressor increases enormously with the blood pressure and presumably the nerve endings in the carotid sinus are much like those in the aorta. We have never tried CO_2 in the cardiac depressor discharge. On the whole, therefore, it is simplest to regard the tone maintained by the carotid sinus as an inhibitory one, the inhibition being lifted by procedures (1), (2) and (3) and increased by (4) and (5).

Rhythm. And so I come to the subject of rhythm, about which I feel that I have very little to say.

I have given reasons for supposing the gasping rhythm to be automatic, and I have shown how such a rhythm may be prolonged into a series of apneuses.

32   "LA FIXITÉ DU MILIEU INTÉRIEUR

The ordinary pneumotaxis rhythm is clearly much more rapid than the gasping rhythm, and I have indicated one way in which such a rhythm might be produced, without very much in the way of new assumptions. In particular I have no explanation to offer of the effect of CO_2 on the rate of respiration, which effect is, as I have said, very inconstant.

Another way, but a way which requires one more assumption, is that the cells responsible for the inhibition are themselves rhythmic, and that the normal rhythm is that of the inhibitions which play upon a gasping centre which always has something in hand and is therefore never permitted to unmask its own rhythm. Here a very pretty analogy could be drawn with the cardiac rhythm, but I hesitate to draw it because after all it is only an analogy, the cardiac mechanism not being of the same specialised nervous nature as the neurones which compose the central nervous system. Indeed, I have perhaps strayed too far into theory or rather speculation. Let me therefore conclude with the enumeration of the specific points which I should like to stress.

Firstly, there are two methods of regulating constancy of the hydrogen-ion concentration, that of evasion and that of correction.

Secondly, the former of these is perfected, in kind, though not in degree, by the time the lower vertebrates are reached.

Thirdly, the latter is common to the whole animal kingdom; but

Fourthly, its most delicate mechanism only develops at the mammalian (and perhaps collaterally at the avian) level.

Fifthly, that most delicate regulation is a regulation by the nervous system, and by a level of the nervous system which is probably higher than the ordinary medullary centres.

REFERENCES

ADRIAN, E. D. and BUYTENDIJK, F. J. J. (1931). *J. Physiol.* **71**, 121.
AHLFELD, J. F. (1890). *Festschrift für Ludwig*, p. 1. Marburg, quoted by Huggett.
ARBORELIUS, M. and LILJESTRAND, G. (1923). *Skand. Arch.* **44**, 233.
BARCROFT, J. and MARGARIA, R. (1931). *J. Physiol.* **72**, 175.
BETHE, A. (1930). *Pflügers Arch.* **224**, 793.
BETHE, A. and WOITAS, E. (1930). *Ibid.* **224**, 824.

CULLEN, G. E. and EARLE, I. P. (1929). *J. Biol. Chem.* **83**, 545.
DAVIES, H. W., HALDANE, J. B. S. and KENNAWAY, E. L. (1920). *J. Physiol.* **54**, 32.
DILL, D. B., TALBOTT, J. H. and EDWARDS, H. T. (1930). *Ibid.* **69**, 267.
ENDRES, G. and TAYLOR, H. (1930). *Proc. Roy. Soc.* B, **107**, 231.
EVANS, C. LOVATT (1919). *J. Physiol.* **53**, 17.
GESELL, R. (1925). *Physiol. Rev.* **5**, 551.
GLASER, R. W. (1925). *J. Gen. Physiol.* **7**, 599.
HALDANE, J. S. and PRIESTLEY, J. G. (1905). *J. Physiol.* **32**, 225.
HARTRIDGE, H. and ROUGHTON, F. J. W. (1925). *Proc. Roy. Soc.* A, **107**, 654.
HENDERSON, L. J. (1928). *Blood.* New Haven.
HERTWIG-HONDRU, L. (1927). *Pflügers Arch.* **216**, 796.
HEYMANS, C. (1931). *C.R. Soc. Biol.* **106**, 34.
HILL, A. V. (1933). *The Science Monthly*, **88**, 316.
HUGGETT, A. ST G. (1930). *J. Physiol.* **69**, 144.
KROGH, A. (1904). *Skand. Arch.* **16**, 348.
LUMSDEN, T. (1923). *J. Physiol.* **57**, 153, 354; **58**, 81, 111.
—— (1924). *Ibid.* **58**, 259.
MARGARIA, R. (1931). Unpublished.
NASH, T. P. and BENEDICT, S. R. (1921). *J. Biol. Chem.* **48**, 463.
PFLÜGER, E. (1868). *Pflügers Arch.* **1**, 82.
PORTER, W. T. (1895). *J. Physiol.* **17**, 455.
REDFIELD, A. C. and FLORKIN, M. (1931). *Biol. Bull.* **61**, 185.
ROHDE, K. (1920). *Pflügers Arch.* **182**, 114.
SVEDBERG, T. (1926). *J. Amer. Chem. Soc.* **48**, 430.
TAYLOR, H. (1930). *J. Physiol.* **69**, 124.
—— (1931). Quoted by Barcroft, J. *J. Hygiene*, **31**, 1.
TAYLOR, H. and TAYLOR, N. B. (1931). *J. Physiol.* **71**, vii.
VAN SLYKE, D. D. (1921). *J. Biol. Chem.* **48**, 153.
ZUNTZ, N. (1877). *Pflügers Arch.* **14**, 619.

CHAPTER II

"LA FIXITÉ DU MILIEU INTÉRIEUR EST LA CONDITION DE LA VIE LIBRE" (*continued*)

TEMPERATURE

Of the conditions which may be investigated the most attractive perhaps is a physical one, namely temperature, and therefore it will be considered in greatest detail. The temperature of man is approximately constant; how did it become so, and with what degree of success did creatures of inconstant temperature perform their bodily functions such as respiration, muscular contraction, heart beat, etc.?

Before these functions are considered in detail a few words may be said in general terms on the implications of alterations of temperature. From the physico-chemical point of view, the activities of the body consist of a vast number of chemical reactions which take place simultaneously. The final result depends upon a complete proportional and quantitative harmony of the rates at which these innumerable chemical reactions are proceeding.

Yet each reaction is a purely quantitative affair and is governed by the general laws of thermodynamics. Each, whether reversible or not, is governed by the principle expressed in the equation of Arrhenius, namely that if the velocities at any two temperatures T_1 and T_2 are respectively k_1 and k_2, and if μ is a constant,

$$k_1 = k_2 e^{\left(\frac{\mu}{2} \times \frac{T_2 - T_1}{T_2 T_1}\right)},$$

the graphical implication of which is that if over a range of temperatures the logarithms of the velocities be plotted as the ordinate, and the reciprocals of the absolute temperatures (on the Kelvin scale) be plotted as the abscissa, the relationship appears as a straight line.

Reactions take place in the body which are not usually regarded as simple, to which the Arrhenius equation still applies. In this connection the equilibrium of haemoglobin with oxygen has been studied with great care. Twenty years ago that reaction was regarded as simple, now it is held to involve the association of four molecules of oxygen with each molecule (mol. wt. 68,000) of haemoglobin as well as the dissociation of the haemoglobin and even the extent to which it is united with sodium. Yet the greater the care with which the determinations are made the more convincing is the rectilinear relation between the logarithm of the

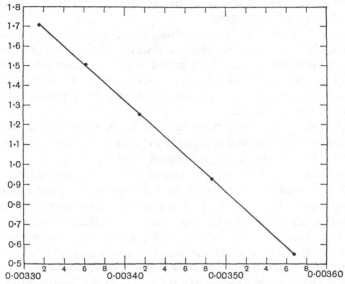

Fig. 19. Ordinate, log. of number proportional to velocity of reduction of haemoglobin. Abscissa, reciprocal of absolute temperature.

50 per cent. saturation pressure of oxygen and the reciprocal of the absolute temperature.

Arrhenius' equation applies to simple chemical reactions, but also without serious error to the velocity of the more complicated reaction which takes place when haemoglobin is deprived of its oxygen by a reducing agent. This has been shown to be the case by Hartridge and Roughton (1925) from whose data Fig. 19 is constructed.

When account is taken of the fact that the velocity of each chemical reaction in the body is simply a property of itself, it is surprising enough that at any one temperature all the reactions of the body should progress at velocities suitable to one another. But it is much more surprising, nay it would seem wellnigh impossible that, granting that the body should function at some one temperature, it should also function over a great range. For if the velocity of some one reaction got out of step with its neighbours the whole machine would jam—to use a phrase which I once heard from the lips of A. V. Hill.

On a casual scrutiny, then, nothing would appear more plausible than to give as a reason for the preservation of a constant temperature, the probable chaos which would result from any considerable thermometric alteration—to say, in short, alter the temperature and the velocities of the reactions which form the physical basis of life must get out of gear.

Yet in reality enquiry shows to what a remarkable extent nature has contrived in one way or another to circumvent any such barrier imposed by this simple application of chemical laws.

How does the body of the animal of variable temperature so control its reactions as to keep them in step over a great range of temperature? In point of fact the poikilotherm seems to have devised methods of overruling to some extent the apparent consequences of the law of logarithmic increase in the rate of velocity of chemical action with temperature. Once we invade the region of living processes there are departures from such a simple rectilinear relation as that shown for the temperature coefficient of the haemoglobin-oxygen equilibrium. Krogh (1916) drew attention to the fact that the metabolism of the body as a whole did not in cold-blooded animals follow the logarithmic law, but fell off relatively to the requirements of that law as the temperature rose. And what is true of the body as a whole seems to be true of such of its individual functions as have been studied. A number of such specialised functions have been collected by Clark (1927), the rate of ciliary movement in *Mytilus* (Gray, 1923), the rate of movement of amoebae (Pantin, 1924), the frequency of the isolated frog's heart (Clark).

Of these, the temperature coefficients over a range of 10° C. do not remain uniform as the temperature rises, but tend to decrease as the temperature increases, thus (Table III):

Table III

Temperature coefficients (Q_{10}) of processes in invertebrate tissues

Temperature ° C.	Frequency of frog's heart	Rate of movement of *Amoebae*	Rate of movement and O_2 consumption of cilia of *Mytilus*
0–10	3·5	7·33	3·1
5–15	2·8	2·71	2·7
10–20	2·4	2·17	2·3
15–25	2·1	—	2·15
20–30	1·8	—	1·95

The above observations have been plotted by Clark; the frequencies in each case being expressed as percentages of that at 20° C.

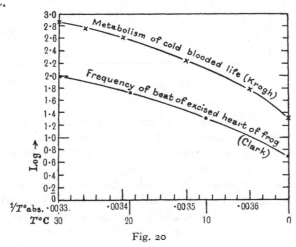

Fig. 20

A closer comparison of some such curves as are shown in Fig. 20 with the data from which they have been drawn leads to considerable doubt as to whether the points observed fall in reality upon a smooth curve. The greater the pains taken to secure accuracy of observation, and the greater the accuracy obtained and the more numerous the observations, the greater becomes the difficulty in drawing satisfactorily any sort of smooth curve through the plotted points which correlate the logarithm of the degree of activity of a vital process with the reciprocal of the absolute temperature at which it takes place. On the other hand Crozier (1924) and his school regard the essential character of such a figure not as being a smooth curve, but as being made up of two or more straight lines, each of which represents faithfully a chemical reaction. Of six

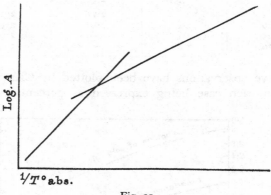

Fig. 21

hundred sets of observations which Crozier has made the majority fall into a simple scheme which involves two chemical reactions, each of which can be represented by a straight line, the lines being disposed more or less as shown in Fig. 21.

It is clear that if we can postulate two such reactions, one of which controls the velocity of the phenomenon observed at one temperature, and another at a higher temperature, we can postulate more than two, in which case a curve of the type shown in Fig. 22 would be obtained.

The numerous phenomena observed cannot, however, all be represented by a series of lines which cross one another after the fashion depicted in Figs. 21 and 22 and the observations fall rather into another scheme, that of a series of parallel straight lines as in Fig. 23.

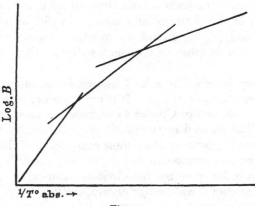

Fig. 22

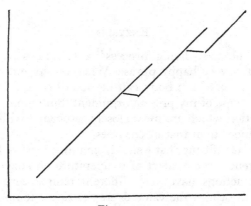

Fig. 23

It is not my purpose here to put forward arguments for or against Crozier's conceptions. Some of the criticism to which they can be subjected is of a very obvious character—some might think it so obvious as to be "cheap". For the rest the reader may be referred to an article in *Biological Reviews* by Bělehrádek (1930).

For my own part I can only say that I started with a prejudice against the scheme which I have sketched out; but having tried to do some experiments with a semblance of accuracy and without any thought of the patterns into which the plotted observations would fall, and having found the points to drop with curious fatality into one or other of Crozier's schemes, the prejudice has passed away.

It is not my purpose here, as I say, either to support or to repudiate Crozier's point of view. It is my purpose, however, to say this: whether you accept Crozier's standpoint or another, it is clear that nature has learned so to exploit the biochemical situation as to escape from the tyranny of a simple application of the Arrhenius equation. She can manipulate living processes in such a way as to rule, and not to be ruled by, the obvious chemical situation. That is true at least over a wide range of temperature.

Having said so much let us pass to the consideration of certain individual phenomena in relation to temperature.

ENZYMES

I used the phrase "living process" a few lines back: possibly that was not a very happy phrase. Whether any individual single chemical process of the body can be described as "living" is far beyond the scope of my present argument, but among the types of chemical action which are most closely associated with life none is more prominent than that of enzymes.

Therefore it is fitting that I should commence what I have to say with a reference to the effect of temperature on enzyme action.

Different actions have very different temperature coefficients, but for the most part the value decreases as the temperature rises. The degree of decrease varies greatly in different cases.

I have plotted some examples which are at once typical and well attested, selected from those given by J. B. S. Haldane (1930).

The point I wish to emphasise in the present connection is that lipase in some way or other has become nearly independent of temperature at the highest temperature at which it has been studied, this being already considerably below that of the homoiothermic animals. How has this occurred? We do not know; yet it seems to be a most interesting and striking case of adaptation by what I have called the method of evasion. Clearly it is not possible to say that the temperature is "buffered", for the actual alteration in temperature takes place. The result, however, is much the same; the

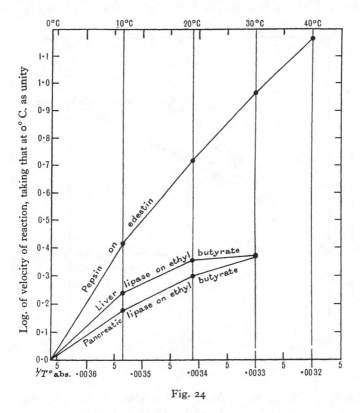

Fig. 24

chemistry of the animal is so twisted that the alteration in temperature imposed upon it has but little effect on the velocity of the reaction. The objection has been raised to some of the above examples, that the medium becomes more acid as the reaction proceeds. This objection may from one point of view be valid;

from another it may provide a hint as to how such things can happen in life.

Intracellular enzymes appear over a considerable range to have rendered themselves independent of temperature, as is shown by R. P. Cook (1930). The author gives three instances, namely the oxidation of $M/150$ formate (see Fig. 25), $M/60$ lactate (see Fig. 26)

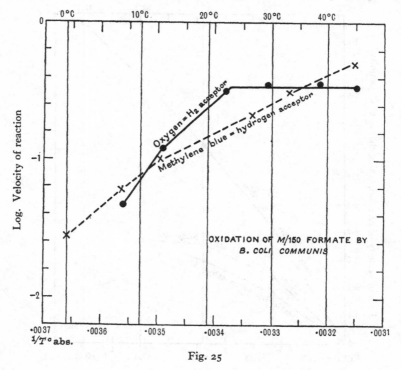

Fig. 25

and $M/60$ succinate by toluene-treated *Bacillus coli* (incidentally we are far enough down in the scale of life) in each case; the oxidation is conceived as a process depending upon the abstraction of hydrogen in a molecule activated by a "dehydrogenase" in the bacillus and the transference of hydrogen to a "hydrogen acceptor", in this case oxygen or methylene blue. One can easily imagine that from the point of view of the bacilli as independent organisms, it is undesirable that these should be limited by the alterations of temperature of their hosts. Is it certain that the enzyme was not reduced in quantity at the higher temperature?

When methylene blue has been used as the acceptor the value of the temperature coefficient falls off rather gradually as the temperature rises, but when oxygen is used, the temperature coefficient

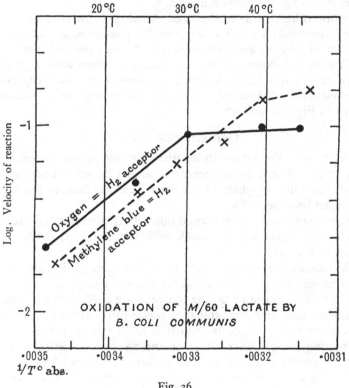

Fig. 26

becomes almost unity, *i.e.* the rate of the enzyme action becomes almost independent of temperature. It is to me extremely interesting that the action in question should be an oxidation, because there are very few chemical actions known which have a temperature coefficient of unity. Of these the reaction which has been most closely studied is also an oxidation and is one fundamental to many forms of life, the union between haemoglobin and oxygen as investigated by Hartridge and Roughton (1925). When I asked Roughton what explanation he could give of so extraordinary a phenomenon he said: "The only one I can think of is that over the range of temperature involved every molecule of oxygen which impinges on the haemoglobin sticks".

On the other hand, in the cases which involve enzyme action there is probably a much more complex phenomenon than in the reaction of oxygen and haemoglobin. As was pointed out to me by Brinkman, a combination of a chemical action with a positive temperature coefficient linked to an adsorptive one with a negative temperature coefficient might produce the type of result which is shown in Figs. 25 and 26.

OXIDATION OF YEAST

We pass from the rather indefinite lowering of the temperature coefficient with rise of temperature, shown by some enzymes, to something which appears much more definite, namely the usage of oxygen by yeast cells.

Four experiments, each carried out with meticulous accuracy by Stier (1932), gave results which tallied perfectly with the type shown in Fig. 21.

The oxidation of yeast may not be at all simple, yet it seems likely to be among the simplest of living phenomena. The results are clear cut. It is certain that the temperature coefficient is lower at the higher temperatures, and it is much more difficult to fit any smooth curve to the points than it is to fit two or perhaps three intersecting straight lines. Figs. 27 and 28 show the effect of temperature on the oxidation of four different cultures of yeast. The results show that the same temperature coefficient is obtainable from culture to culture.

THE HEART BEAT

With the object of ascertaining something about the factors which were involved in the regulation of the heart beat in relation to temperature, Izquierdo and I (1931, a) carried out some experiments on

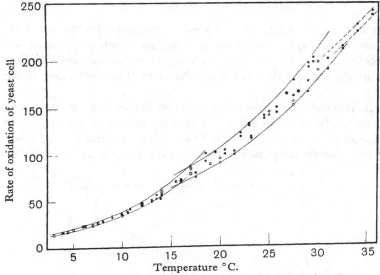

Fig. 27. Rate of consumption of oxygen by yeast plotted against temperature in °C.

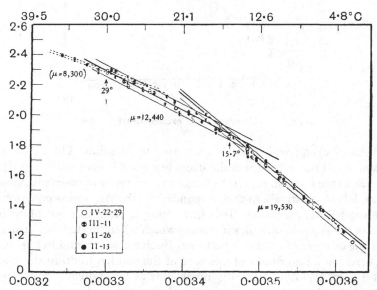

Fig. 28. Logarithm of rate of consumption of oxygen by yeast plotted against reciprocal of absolute temperature.

the heart of the frog. The plan was to compare the effect of temperature on: (*a*) the rate of the perfused heart (which of course was necessarily free from immediate humoral or nervous control from without), and (*b*) the rate of the heart in the intact body of the frog.

It happened that one batch of experiments was carried out in January, the other in June. In January on the excised heart the relation between the logarithm of the pulse rate and the reciprocal of the absolute temperature was a linear one up to 20° C. or there-

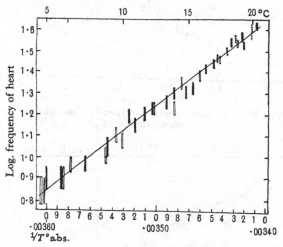

Fig. 29. Common frog, excised heart, winter.

abouts, beyond which the heart ceased to function. The machine jammed. That at least was the usual but not the invariable finding. Such a case is illustrated in Fig. 29. In a minority of cases the pulse rate fell off relatively to that demanded by the Arrhenius equation, as the temperature rose. It is interesting to note that out of such cases it was possible to pick instances which when plotted conformed to one or other of Crozier's patterns. Such a case is found in Fig. 30. There can be no doubt in this case of the sudden jump in the line at 15–17° C. But need this break always be sudden? Such instances as shown in Fig. 31 could easily be explained on the assumption that the displacement of the upper part of the curve was not sudden but gradual (Fig. 32).

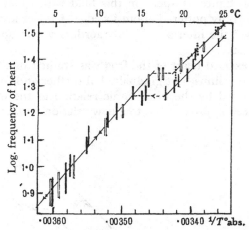

Fig. 30. Common frog, excised heart, winter.

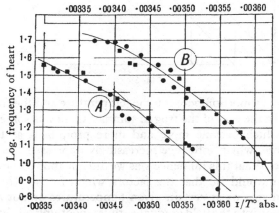

Fig. 31. Common frog, excised heart, winter.

The fact that we have obtained all Crozier's patterns from the beat of the frog's excised heart seems to indicate that the difference between one pattern and another is not fundamental, and it is clear that all the patterns could be obtained on the supposition that the cause of the break in Fig. 30 need not necessarily be so sudden as shown in that figure. I speak in this field with great diffidence, because I have only pottered about its edge. Crozier and his school have developed the interior with extraordinary care and beautiful experimentation.

When the excised heart of the frog was studied in summer, the frequency of the sinus beat gradually fell further and further short of that demanded by the logarithmic relation as the temperature rose, and indeed approximated to a new relation, namely a simpler

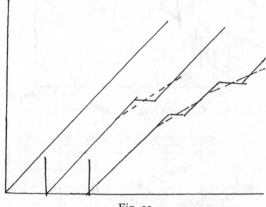

Fig. 32

arithmetical proportion between the temperature and the frequency (Fig. 33). In the case of the frog in summer there was some gradual influence which became more and more prominent as the temperature rose, which influence increasingly depressed the velocity of the pulse. What this influence was we did not know. The simplest assumption would be that the factors which in the winter time operated in spasms at certain given temperatures or not at all, in summer came into the field gradually, thus smoothing off the curves. One day Crozier will tell us whether or not this is so—my immediate point is another, namely to emphasise the fact that this departure from the law of Arrhenius is shown in the excised heart and not only in the intact animal. It was in the intact animal that we first observed it, and it would have seemed reasonable to suppose that the damping of the heart rate with temperature was connected with the very intactness of the organism, that, for instance, it was due to vagus impulses coming into the field or to the secretion of some hormone. Such an explanation might seem to be ruled out by the fact that the curves which relate the temperature and the heart frequency are practically identical whether the object of

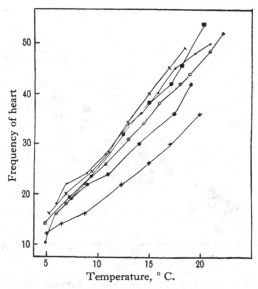

Fig. 33. Common frog, excised heart. Summer.

study be (1) the isolated heart, (2) that of the normal intact frog, and (3) that of the atropinised intact frog (Fig. 34). The recent work of Carter (1933) however has proved it to be otherwise. Carter has confirmed the general difference in the shape of the temperature-pulse curve as between the summer and winter frogs, and he has added some points of great interest.

1. The summer (linear) curve can be altered towards the winter (exponential) curve by immersion of the heart for sufficient time in Ringer's solution.

2. The winter (exponential) curve can be altered to the summer (linear) one by treatment with thyroxine.

3. Apart from much larger doses thyroxine is specific.

4. Thyroxine is antagonised by extract of the anterior lobe of the hypophysis or by bromides.

In Carter's view the pulse rate of the summer frog is damped down at higher temperatures by thyroxine, presumably acquired,

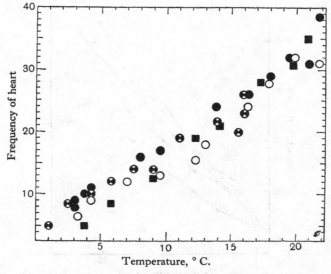

Fig. 34. Common frog, heart in intact animal. Summer: ☉ going down, ○ going up. Atropinised: ● going down, ■ going up.

during life, from the blood stream and adsorbed by the cardiac tissue.

The range of temperature over which the heart of the common frog beats (about 20° C.) is all too small to bring out the distinction between the logarithmic and the linear relation between temperature and frequency, another 10° would make a great difference. In this respect we had a stroke of good fortune. Prof. Hogben, then stationed in Cape Town, came into the laboratory and, sympathetic as ever, he pointed out that the South African clawed toad (*Xenopus laevis*) could withstand change of temperature of something approaching 40° C. as had also been found by Clark. Not only so but he very kindly sent in a consignment of these toads.

Fig. 35 shows the comparison (made by N. B. Taylor, 1931) between the excised heart of the common frog (*Rana temporaria*) and that of *Xenopus*. Over the range covered by the former, the curves given by the two species agree in type and in fact almost coincide; the curve from *Xenopus*, however, continues in its linear

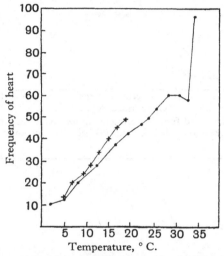

Fig. 35. Comparison of typical temperature-excised heart rate curves of British frog and South African toad respectively. + British; · South African. (N. B. Taylor.)

direction up to almost 30° C., then it falls off, so that at 33° C. the rate is less than at 29° C. If the temperature be further raised there is an abrupt increase in the sinus rate followed by death. The last phase must be regarded as beyond the region of function because it is irreversible. It appears to be due to a circus movement, and once it has supervened death is in any case only a matter of a short time.

In what may be regarded in the excised heart as "physiological", *i.e.* between 1° and 30° C. or thereabouts (differing by a degree or two in different hearts), the pulse rate bears a roughly linear relation to the temperature and not a logarithmic relation.

And now to pass to the heart in the intact animal. There was this great difference between *Xenopus* and *Rana*, namely, that the vagus exerts a marked action in *Xenopus*. This action appears to be almost

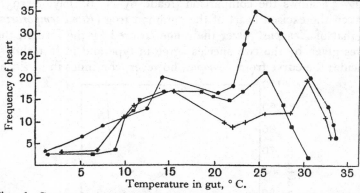

Fig. 36. Curves obtained from experiments upon intact heart of South African toad. Temperatures recorded by means of thermocouple in the bowel. (N. B. Taylor.)

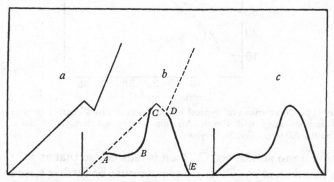

Fig. 37. *A, B, C* = vagus effect. *D, E* = period of death without circus movement.

absent at low temperatures, at its maximum at moderate temperatures, and it falls off at high ones (Fig. 36).

As compared with the excised heart, that in the intact animal shows another departure, the rapid rise of rate before death has never been obtained.

Fig. 37 shows schematically the relation between the curves obtained from the excised (*a*) and intact heart (*c*) and the reasons to which the differences between the two may be attributed.

This in fact is the type of curve given by the heart of the intact *Xenopus*. The beats were counted by the string galvanometer, and the temperature measured in the bowel.

Yet the matter cannot be disposed of quite so simply. If so the atropinised intact *Xenopus* heart should give the same curve or nearly the same as the excised heart. That was so with the heart of the common frog. In the few experiments which have been carried out on *Xenopus* the atropinised heart gives a curve, intermediate between that of the normal heart and of the excised heart. The suggestion of two summits remains.

The natural transition from the lower vertebrates to the mammalia is by way of hibernating mammals. In the present connection the marmot has been, of these, the most completely studied.

The heart of the marmot will beat through the same range of temperature as has been discussed for the frog; the result for the perfused marmot's heart (as found by Endres, Matthews, Taylor and Dale, 1930), however, differs from that obtained for the perfused frog's heart. If the perfusion was started at 28° C. and the temperature gradually lowered, the heart beat, normal in character, became slower until about 17° C. was reached. Over that range the rate of beat had conformed to the logarithmic law and showed a temperature coefficient of almost exactly 2. Between 17° and 16° C. the heart beat suddenly became much slower (see Figs. 38 and 39). The reason was at once revealed by the string galvanometer record, the *P* wave had disappeared from its normal position and the ventricle was beating with a new and slower pacemaker.

In the marmot—again differing from the common frog—the reaction to temperature of the heart of the intact animal differed greatly from the reactions of the perfused heart.

In the short series of experiments performed in Cambridge lately, it early became apparent that special arrangements were required for the registration of temperature; it was therefore recorded both in the heart itself and in the rectum. At present we are concerned only with the temperature of the heart itself. The experiment now to be described commenced with a heart temperature of

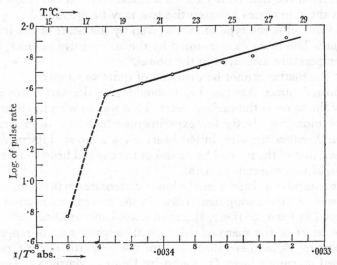

Fig. 38. Excised heart of marmot. Logarithm of heart rate plotted against reciprocal of absolute temperature. (By permission of the Royal Society.)

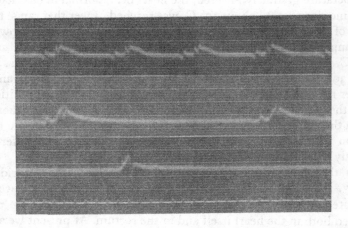

Fig. 39. Electrocardiogram of excised and perfused heart of marmot at 25·2°, 17° and 16° C. respectively. (By permission of the Royal Society.)

about 10° C. At this temperature the heart was irregular, about six beats per minute on the average, and the electrocardiogram did not indicate a definite P wave. As the temperature gradually rose the heart steadied, and at 13° C. the heart was beating regularly and with a normal electrocardiogram. At about 14° C., however, the whole picture altered, and by 17° C. the heart beat rose to 100. It

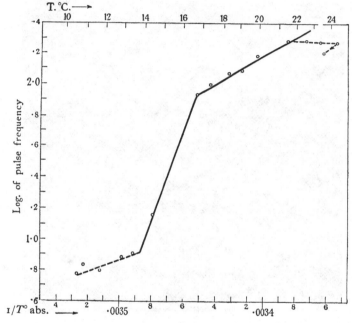

Fig. 40. Heart of intact marmot. Relation of log. of frequency to reciprocal of absolute temperature. (Upper scale = temperature in ° C.) (By permission of the Royal Society.)

was clear that something catastrophic had taken place (Fig. 40). The marmot was awakening.

There is no real analogy between the sudden rise shown in Fig. 40 and that shown in Fig. 38, because electrocardiograms taken at temperatures of 10°–12° C. show well-marked P waves (Fig. 41). Indeed it might be contended with some show of reason, though I think it could not be said with certainty, that the P wave

persists at a lower temperature than do the ventricular waves
(Fig. 41 c).

No; a more probable explanation of the sudden rise shown in
Fig. 40 is along lines suggested by the work of Gellhorn (1924).
Gellhorn shows that adrenaline increases the temperature coefficient

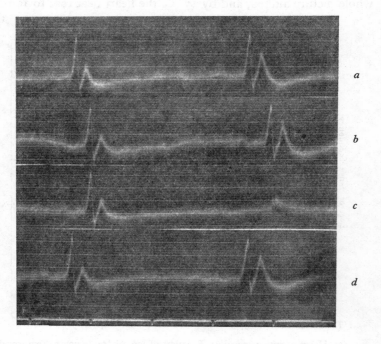

Fig. 41, a–d. Four consecutive pairs of heart beats, which take place at quarter-
minute intervals. Time marker, 1 second. (By permission of the Royal Society.)

of the frequency of the heart. Clearly an outpouring of adrenaline
would at once produce the sort of sudden rise which occurs when
the marmot awakes and would also account for the higher value for
the temperature coefficient in region 16°–21° C. as compared with
that at 11°–14° C. Such a pouring out of adrenaline is not merely
a manifestation of the activity of the suprarenal medulla, behind

that are stimuli arriving along the splanchnics and behind them is the brain.

Evidently the whole venue had changed; no longer were we studying the effect of the temperature on the heart beat as such, but the effect of temperature on some more complex system (see Carter) and here at the threshold of homoiothermism we encounter the sympathetic system as a factor in the adaptation of the heart rate to temperature.

The awakening of the marmot seems primarily to involve quite a high level of the brain; the mere difference between wakefulness and sleep according to Hess (1929) is a manifestation of altered activity of certain centres in the brain, whilst the most obvious signs of the awakening in the marmot are intense shivering and the invocation of the whole mechanism for the augmentation of heat production. In this mechanism acceleration of the heart is merely an incident.

Attention has been drawn to the increase in the value of the temperature coefficient in the region around 20° C. If the temperature coefficient in that region were maintained the pulse would soon be racing, but another factor enters, namely, the vagus. At 23° C. (Fig. 42 a) the pulse rate is uniform, at 26° C. (Fig. 42 b) signs of arrhythmia are seen, whilst in Fig. 43, at 29° C., the arrhythmia is of the most pronounced type.

The contrast between the effects of temperature on the heart, when the heart is and is not harnessed to the thermogenic centre (if I may use the term), is well shown in the cat. As is well known, animals when deeply narcotised lose their power of heat regulation and become in effect poikilothermic.

If the deeply narcotised (luminal or chloralose) cat be cooled the rate of heart beat falls progressively with the fall of temperature and the cat remains quiescent. It behaves much as the frog would do under similar circumstances. The velocity of the pulse always varies in the same sense as the temperature of the cat. Indeed, like the heart beat of the frog and like the excised mammalian heart (Knowlton and Starling, 1912), the pulse is, from 33° to 44° C., roughly a linear function of the temperature (Fig. 44 a) (Barcroft and Izquierdo, 1931 b).

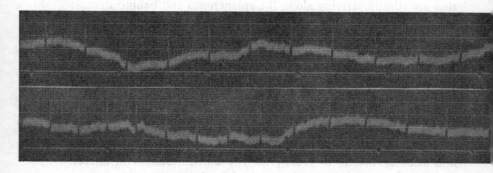

Fig. 42, *a* and *b*. Electrocardiograms at 23° C. and 26° C. respectively; the former shows no arrhythmia, the latter shows commencing arrhythmia. Time marker, 1 second. (By permission of the Royal Society.)

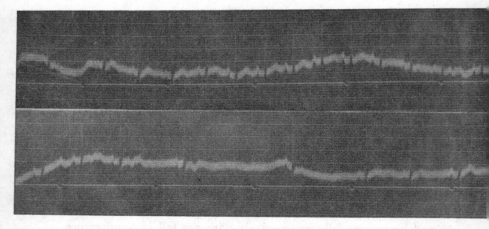

Fig. 43. The upper and lower portions read continuously. Same experiment as Fig. 42, 29·10° C. Arrhythmia well established. Time marker, 1 second. (By permission of the Royal Society.)

If on the other hand the cat be narcotised very lightly (no more than sufficient to make the handling of the animal possible), quite a different picture presents itself. The pulse, as the animal is cooled below the normal body temperature, tends to rise instead of fall,

and that tendency to rise is maintained over a drop in temperature
of several degrees.

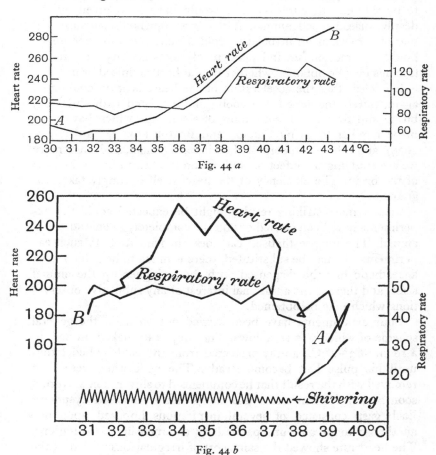

Fig. 44 *a*, from deeply anaesthetised, and 44 *b* from lightly anaesthetised cat,
ordinate heart and respiration rates, abscissa temperature ° C.

Fig. 44 *b* shows the mean pulse rate, but it is scarcely fair to
speak of a mean pulse rate in the region immediately below 37°,
for the pulse rate fluctuates momentarily in a quite unusual

and abnormal way. The line of maximum pulse rate represents a summit, the line of minimum pulse rate at any temperature seems to be the smooth curve such as would have been given by the deeply anaesthetised animal. A new phenomenon then faces us, that of a mechanism invoked by cold which accelerates the pulse, but it is a mechanism independent of vagus activity, for similar tracings may be obtained whether the cat be atropinised or not.

It is clear that the acceleration of the heart beat on cooling is, again, purely incidental, the cooling is associated with shivering, biting and the general symptoms of sham rage which have been made familiar to us by Cannon and Britton (1925), Bard (1928, 1929), and other members of the Harvard school, and once more we are studying the effect of cold not on the heart but on the base of the brain. The efficiency of the heart itself is simply taken for granted.

Clearly more striking results might be expected could the experiments have been carried out on the completely unanaesthetised animal. The unanaesthetised cat does not lend itself to such experiments. If man be substituted, there is much to be gained. The anaesthetic may be dispensed with, he is higher up the animal scale, and there is the additional interest of any subjective observations which may be obtained.

Four experiments have been carried out on man, the general scheme of which was as follows. The subject lay naked, in bed, in a room at 3°–4° C., amply protected from the cold by bedclothes until his pulse had become steady. The bedclothes were then removed with the result that he commenced to shiver. The shivering soon became violent and was accompanied by marked dyspnoea. Each gasp consisted of several inspirations imposed upon one another without and unseparated by commensurate expirations. The heart rate showed the same sort of irregularities as that of the cat but on a much more striking scale (Fig. 45), for the whole drop in rectal temperature over about half an hour was less than 1° C.

When the bedclothes were replaced and without any rise in body temperature the shivering passed off and the pulse dropped to a level below that observed after it became stabilised at the commencement of the experiment. The experiments on man yielded in a more definite form what those on the cat had rather indicated, that the rise of pulse rate which takes place on exposure was part of the same syndrome as the shivering. Moreover, confirming the results of Magnus and Liljestrand (1922), it appeared that the

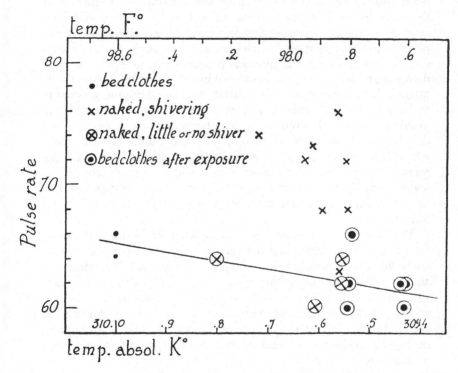

Fig. 45. Relation of rectal temperature to pulse rate in man.

shivering was itself due to a reflex from the skin, for as soon as the skin was warmed the shivering ceased.

It seems possible to couple these experiments with those of Cannon and Britton (1925), who induced shivering and a rapid pulse rate in cats by placing ice-cold water in the stomach, or by placing the cats in a draught. Unfortunately the body temperature was not recorded in these experiments, but the point which appeared was that the results depended upon the intactness of the sympathetic system, for they were not obtained in animals deprived of it.

To sum up, therefore, what we have said about the effect of temperature on the heart is this: That in the frog in winter the excised heart usually obeys the logarithmic law relating its frequency to the temperature; in exceptional cases curves may be obtained showing Crozier's patterns; that in summer it is subjected to temperatures so high that the logarithmic relation would enforce much too high a heart rate; that by some process the frequency has been damped at higher temperatures, making the relation between frequency and temperature nearly linear; that the damping process is to be seen in both the excised and the "intact" heart and has nothing to do with vagus or sympathetic action but depends on the action of adsorbed thyroxine. In the South African toad, which has a much larger range of temperature, the basal arrangement appears to be quite similar to that of the common frog (*Rana temporaria*) in summer, but in addition the vagus comes into the field and slows the heart even more than otherwise would have been the case.

The heart of the hibernating marmot when excised and perfused follows the logarithmic law between certain temperatures. Above about 30° C. it "jams", below about 15° C. the heart becomes very slow, a nodal beat only taking place. In the living animal the heart has a range of temperature from about 1° to about 30° C. The relation of the logarithm of the frequency to $1/T$ is linear, but with a great break at about 15° C., where the heart suddenly accelerates (at higher temperatures acceleration is much less and there is very pronounced vagus action).

In the deeply narcotised cat there is a rise of pulse rate with rise of temperature, but if the narcosis be light the fall of temperature for some degrees below the normal body temperature produces a rise in pulse rate, which apparently is associated with shivering and depends on a nervous factor superposed upon and overruling the more fundamental relation seen in cold-blooded animals.

Here then is a very fine issue—the cold-blooded animal successfully adopting ingenious mechanisms, first biochemical, then physiological, in order to adapt its heart to the variations of its environment;

the warm-blooded animal discarding what its cold-blooded predecessor has laboriously beaten out, invoking the nervous system
to reverse the normal biochemical relationship and gaining a new
freedom by adapting, not itself to the internal environment, but
the internal environment to itself.

RESPIRATION

Our consideration of the effect of temperature on the heart beat has
quite failed to show that a constant temperature is of any advantage
to the heart; on the other hand, that organ can function quite
satisfactorily over a great range of temperature. Moreover, at the
point at which temperature regulation effects a homoiothermic state
the heart becomes a mere servant, its own interests being apparently
taken for granted.

While respiration is naturally associated in the mind with the
thorax and even with the heart it is constantly necessary to recollect
that essentially respiration is conducted in the brain. Our study of
the effect of temperature on respiration is the study of the reaction
of cephalic—not thoracic—processes to environment. That indeed
adds additional interest. For while it might reasonably be held that
the organs of the body had, before the evolution of the thermotaxic
mechanism, been so evolved as to be efficient over a great range
of temperatures, it might equally be held that once the brain is
touched constancy of environment is all important, and that
"la vie libre" involves the whole of the efficiency of cephalic
processes.

Yet the study of the effect of temperature on respiration leads
us into a room in which the pictures are curiously similar to those
which we have just left.

We may start as low down the animal kingdom as we please.
Crozier and Stier (1925) have determined the temperature coefficients of the respiratory movements of various forms of insect
life, and have applied their methods to the observations of other
workers.

Here I need only refer to two sets of readings, one on the larva of
Anax which gives a straight line when the logarithm of the fre-

quency is plotted against the reciprocal of the absolute temperature, the second being on a decapitated grasshopper. This latter shows one of the characteristic patterns which we have already seen in heart tracings. But in each case the respiration is capable of accommodating itself to temperatures which cover a considerable range, in the case of the grasshopper 18°–40° C. Our interest, however, is more directly with the vertebrates, and therefore we may turn to the respiratory rhythm of the goldfish, which is of special interest in view of the observations of Adrian and Buytendijk (1931), to which allusion has been made in a previous section. Many workers have compared the frequency of the opercular movements with the alterations in the temperature of the water. Crozier and Stier, who give a comprehensive account of the subject, show that over a range of about 20° C., from about 8°–26° C., the frequency of the opercular movements obeys the logarithmic law and varies with temperature in precisely the same way as the oxygen consumption of the fish.

Passing from the goldfish to the frog, Crozier and Stier have plotted the relation between pharyngeal movements and the temperature, again plotting the logarithms of the frequency against the reciprocal of the absolute temperature. In preparations the forebrain of which is destroyed there is at least the suggestion of a break, at about 20° C. ($1/T°$abs. 0·00341), of exactly the same nature as the break which we have seen in the corresponding curve for the heart.

Before passing to the mammalia I should perhaps point out that while the gill movement of the fish may morphologically be the parent of the gasp, the pharyngeal movement of the frog occupies a rather ambiguous position. It is not admitted by Crozier and Stier that the pharyngeal movement is, as is generally supposed, the true respiratory movement. They regard the movement of the throat as serving merely to renew air within the buccal cavity while the glottis is closed, and as "not proper movements of lung ventilation (Baglioni, 1900, 1911), although presumably innervated from the respiratory centre (for review of innervation see Black, 1917). The flank movements, indicating lung filling, seem, however, to follow the same curve so far as our observation extended".

In the deeply narcotised cat (Barcroft and Izquierdo, 1931 b) the rate of respiration bears a rough logarithmic relation to the reciprocal of the absolute temperature between the temperatures of 31° and 41° C. and therefore bears the sort of relation which might be expected in a cold-blooded animal (see Fig. 46). Each respiration appears to be normal in kind, and it does not seem possible to say that the mechanism of respiration judged as such is at a disadvantage. But when the narcosis is very light the whole picture is altered, as it was in the case of the heart beat (see Fig. 47).

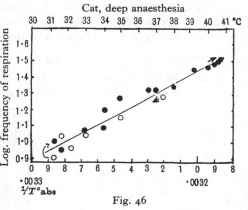

Fig. 46

The cardiac rhythm, we saw, became an incident in the mechanism of heat production. The connection between the respiratory rhythm and the thermotaxic mechanism is even more intimate. Indeed, in animals which do not sweat respiration is not merely connected with the machinery for temperature regulation, it *becomes*

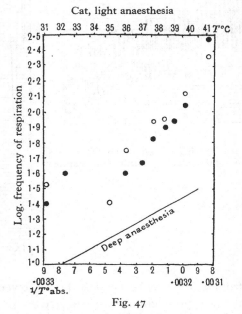

Fig. 47

in great part the machinery for temperature regulation. That is a beautiful inversion; the mechanism for the evasion turned upside down and turned into the mechanism for correction.

What then is the relation of respiration to heat regulation? This may be considered with regard to displacement of temperature on either side of the normal, first in relation to the lowering of temperature and secondly in relation to rise of temperature.

Relation of respiration to lowering of temperature in man

In the experiments of which I have already spoken in which the human body was exposed to a low temperature, tracings were taken

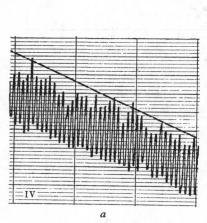

a

Fig. 48. Tracing of respiration taken with Dubois' spirometer: *a*, normal temperature; *b*, 36·6° C.

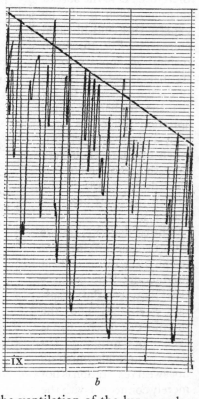

b

from which may be deduced both the ventilation of the lungs and the heat production of the subject.

Fig. 48 *a* shows the course of respiration at the normal temperature, and Fig. 48 *b* shows the same when the temperature was reduced to 36·6° C. The normal sequence of inspiration and expiration is abolished and is replaced by a series of very deep inspirations. On closer scrutiny it seems a little inaccurate to call them deep inspirations. The upstroke in each case is a series of inspiratory efforts, the expirations between them having disappeared. The upstroke is in fact just like that of a cyanide apneusis, and the breath is held for an appreciable time before the apneusis breaks down. We are dealing with a region of the brain at least as high as that which differentiates between pneumotaxis and apneusis. But that is not the whole story, because, as Dr W. G. Garrey pointed out in conversation beforehand, and observation confirmed him, the shivering fits with which respiration of this type is associated always occur on the inspiratory phase. The two are coupled together, suggesting that either inspiration causes shivering or shivering causes inspiration: of the two the former seems the more probable, namely that the activity of the inspiratory centre is allowed to irradiate in some way over the central connection of the muscles involved in shivering in much the same way as it irradiates over the centres which govern muscles of forced respiration to form a gasp. If it is not more than a slight exaggeration to say that a rigor is an apneusis run riot, it is not more than a slight exaggeration to say that the regulation of respiration becomes the regulation of heat production when the temperature is artificially lowered.

Relation between respiration and rise of temperature

When the temperature is raised in man the principal vehicle of heat loss is the evaporation of sweat. In most of the higher animals this is not the case. In them heat loss is effected largely by the evaporation of water from the mouth and tongue and respiratory passages. The amount of heat lost depends upon the quantity of water evaporated, which in turn is a function of the volume of air which traverses the respiratory system. Nevertheless, it is clear that ventilation of the lungs must be ruled by quite other considerations, namely the preservation of the constancy of CO_2 in the alveolar air, which in its turn depends upon the metabolism. The organism is therefore presented with the issue of how to reconcile two opposing elements: (1) in order to diminish heat production the metabolism should be reduced to the minimum with consequent reduction of the alveolar ventilation; (2) to promote heat loss there should be a maximum of evaporation with a consequent proportional increase in total ventilation. The body has found a solution in the phenomenon known as tachypnoea which you may see any day when a dog lies down in the warm sunshine—rapid respiration, but so shallow that the alveoli are little ventilated, whilst the amount of air which passes over the mucosa of the mouth, the windpipe and the bronchi is very great. The opposite combination of events occurs when the body is cooled, namely an increase of the metabolism and so of the CO_2 elimination whilst reducing the total ventilation to the minimum compatible with the preservation of a constant alveolar CO_2. The breath is then held for a while and exhaled. The total ventilation is increased, as it is in tachypnoea, but note that the process is of precisely the opposite type as regards evaporation. As the result of these large respiratory spasms the ratio of alveolar ventilation to dead-space ventilation is at its maximum, not at its minimum, and therefore there is the least degree of cooling for a given oxygen uptake. And so we are faced in most mammals (Equidae excepted) with a co-ordinated machine by which a mechanism evolved for the purpose of oxygenating the blood is pressed into service for the regulation both of hydrogen ions and temperature, and is so manipulated as to do the work whether these variables alter in the same or in opposite senses. Judged as a *tour de force* this mechanism must be almost without a rival, but it is not surprising that in man, the form of mammal which demands the finest adjustment both of temperature and of hydrogen ions, such a triple purpose mechanism, however ingenious, should be discarded and an installation provided for the independent regulation of temperature. This installation is the highly developed functional skin with its deep layer of material which is very retentive of heat when the vessels are constricted and its power of copious secretion designed to "eliminate" calories on the great scale.

The last point to stress is one which will not have escaped the reader, namely that the actual fight for the preservation of a constant internal medium is carried out in the brain—and if

Barbour and Prince (1914) are correct, rather high up in the brain, namely in the corpus striatum. The mere locality in which the seat of exact temperature regulation is situated suggests that this is a development of the higher vertebrate types.

Our review in this section started with the enzymes which were on the borderland of life, we saw in them a mechanism which was of the nature of buffering, an acceptance of temperature variation and an adaptation of the organism to it, the method of "evasion", and we have ended by the reversion of the process, the refusal to accept alteration of temperature, and the domination of the method of elimination.

We started with a purely chemical phenomenon in a unicellular organism, we end somewhere just below the cerebral hemispheres.

REFERENCES

ADRIAN, E. D. and BUYTENDIJK, F. J. J. (1931). *J. Physiol.* **71**, 121.

BAGLIONI, S. (1900). *Arch. Anat. Physiol.* Suppl. **33**.

—— (1911). *Ergeb. Physiol.* **11**, 531.

BARBOUR, H. G. and PRINCE, A. L. (1914). *J. Pharm. Exp. Therap.* **6**, 1.

BARCROFT, J. and IZQUIERDO, J. J. (1931 *a*). *J. Physiol.* **71**, 145.

—— —— (1931 *b*). *Ibid.* **71**, 364.

BARD, P. (1928). *Amer. J. Physiol.* **84**, 491.

—— (1929). *Arch. Neur. Psy.* **22**, 230.

BĚLEHRÁDEK, J. (1930). *Biol. Rev.* **5**, 30.

BLACK, D. (1917). *J. Comp. Neurol.* **28**, 379.

CANNON, W. B. and BRITTON, S. W. (1925). *Amer. J. Physiol.* **72**, 233.

CARTER, G. S. (1933). *Journ. Exp. Biol.* **10**, 256.

CLARK, A. J. (1927). *The Comparative Physiology of the Heart*, p. 65. Cambridge.

COOK, R. P. (1930). *Biochem. J.* **24**, 1538.

CROZIER, W. J. (1924). *J. Gen. Physiol.* **7**, 189.

CROZIER, W. J. and STIER, T. B. (1925). *Ibid.* **7**, 429.

ENDRES, G., MATTHEWS, B. H. C., TAYLOR, H. and DALE, A. (1930). *Proc. Roy. Soc.* B, **107**, 222.

GELLHORN, E. (1924). *Pflügers Arch.* **203**, 163.

GRAY, J. (1923). *Proc. Roy. Soc.* B, **95**, 6.

HALDANE, J. B. S. (1930). *Enzymes.* London.

HARTRIDGE, H. and ROUGHTON, F. J. W. (1925). *Proc. Roy. Soc.* A, **107**, 654.

HESS, W. R. (1929). *Amer. J. Physiol.* **40**, 386.
KNOWLTON, F. P. and STARLING, E. H. (1912). *J. Physiol.* **44**, 206.
KROGH, A. (1916). *The Respiratory Exchange of Animals and Man*, p. 98. London.
MAGNUS, R. and LILJESTRAND, G. (1922). *Pflüger's Arch.* **193**, 527.
PANTIN, C. F. A. (1924). *Brit. J. Exp. Biol.* **1**, 519.
STIER, T. B. (1932). Unpublished.
TAYLOR, N. B. (1931). *J. Physiol.* **71**, 156.

CHAPTER III

"LA FIXITÉ DU MILIEU INTÉRIEUR EST LA CONDITION DE LA VIE LIBRE" (*continued*)

OXYGEN

When Claude Bernard put forward the principle of constancy of internal environment he instanced three substances, one of which was oxygen.

The mention of oxygen raises a point which hitherto has not been discussed. The environment of the cell is not really blood but lymph, blood as an approximation merely. If the amount of any particular material in the blood is large, and if the amount lost during the passage of the blood through the capillary is small, the composition of the blood plasma may be taken as being a sufficiently close approximation to that of the lymph. In the case of oxygen, however, these conditions do not hold good. The blood loses a quarter to a third of its oxygen in passing along the capillary. What then is the closest approximation to the internal environment of this cell which can be expressed in terms of the composition of the plasma? Clearly the plasma of the venous rather than the arterial blood, for with it the lymph is more nearly in equilibrium.

At first sight it may seem that the oxygen content of the lymph would be more constant if it were governed by the oxygen in the arterial rather than the oxygen in the venous blood. The arterial blood is relatively constant in relation to its oxygen content, and the venous blood is relatively variable.

Points arise, however, to modify this conception: in the first place it is the concentration of oxygen in the *plasma* which counts. The haemoglobin is practically an oxygen buffer of a very complete kind and the oxygen dissociation curve may be regarded much in the light of an oxygen titration curve. By one of the miracles of nature it has the general properties of other curves which are ob tained by the titration of buffered solutions, namely at the extreme ends of the curve a small addition of material buffered makes a large alteration in its potential, but towards the centre of the curve a large addition of the material buffered makes but a small alteration in potential.

At no point on the curve is the oxygen pressure (*i.e.* the concentration in the plasma) independent of the quantity of oxygen in the blood, but in ordinary oxygenated blood in the region situated symmetrically about that of half-saturation the blood can impart or take up over 40 per cent. of its possible oxygen content within an extreme range of pressure of 20 mm. Hg. But that is not all, for in the mammals carbonic acid exchange takes place approximately synchronously with oxygen exchange. As was shown by Christiansen, Douglas and Haldane (1914) the amount of oxygen which the blood uses for a given alteration of oxygen pressure is increased by the fact of its acquiring carbonic acid in nearly equivalent quantities. Therefore when we consider the dissociation curve of the arterial blood, and even more emphatically is it true of the circulating blood, very different degrees of unloading are consistent with very small differences in pressure (Fig. 49). In Bock, for instance, the saturation of the venous blood varied between 77 per cent. and 41 per cent. On the curve for the circulating blood given by Christiansen, Douglas and Haldane, the difference of saturations would amount to about 15 mm. oxygen pressure. But here comes another point: when heavy exercise is taken such as would reduce the saturation of the venous blood to 40 per cent. the blood alters in reaction, the dissociated curve therefore shifts. This shift, as shown by the Monte Rosa Expedition of 1911, may result in as much as 7 mm. increase in oxygen pressure at 40 per cent. saturation. It is quite possible, therefore, that between such slight exercise as would admit of the venous blood being 75 per cent. saturated, and such heavy exercise as would reduce its saturation to 40 per cent., there is a pressure difference in the mixed venous blood of less than 10 mm. of mercury.

To pass to the evolution of haemoglobin. It used to be said that haemoglobin had been evolved two or three times over. The statement has always seemed to me to be a little vague. I suppose it meant that haemoglobin had cropped up, apparently spontaneously, in several phyla of the animal kingdom which were not in the same line of descent. However that may be, the whole subject has taken on an entirely new aspect within the last six or seven years.

This progress has been due to the work of Keilin (1925, 1926), who originally was interested in the problem of whether the haemoglobin of the parasitic larva of *Gastrophilus* is or is not the same as that of the host—namely the horse—in which it develops.

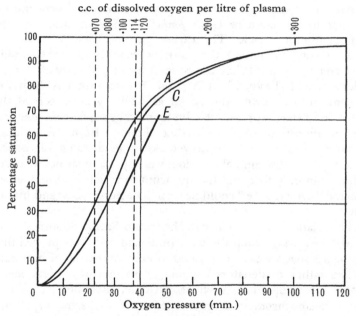

Fig. 49. *A* = arterial blood; *C* = circulating blood (rest); *E* = circulating blood (exercise).

Gastrophilus, a fly, and the horse are far enough removed phylogenetically, but is their haemoglobin the same, and if so is it merely the haemoglobin of the horse that is laid down in the tissues of the larva?

The two are spectroscopically different. The point at once arose, what happened to the haemoglobin of the larva, did it persist into the adult? This question led to an inspection of the muscles of the fly in which was discovered, not the spectrum of haemoglobin, but the now well-known spectrum of cytochrome. The enquiry then spread to the muscles of butterflies and other forms which had no connection with blood as a form of food and finally to such bodies as yeast.

Since the haemoglobin laid down in the tissue of an animal is different from that found in its blood, the more interesting comparison would be between the haemoglobin in horse muscle and that in the larva of *Gastrophilus*. Hence an attack on the haemoglobin in muscle. The earlier part of Keilin's work was largely a confirmation of the work of MacMunn (1886, 1889). The latter when published ran counter to the authority of Hoppe-Seyler's school (Levy, 1889), and on that account largely it did not come into its own. But so far as Keilin was concerned the muscle was a starting-point. Keilin found that a spectrum identical with, or almost identical with, that obtained from the reduced muscular tissue was to be observed over a great range of tissues. These tissues in the animal kingdom were spread over nearly every phylum: not only that, but the spectrum to which Keilin attached the name "cytochrome" could be seen in bacteria, yeast and higher plants.

What relation, if any, has cytochrome to haemoglobin? Anson and Mirsky (1925), about the same time, had definitely proved that haemochromogen was a conjugated protein consisting of haematin attached either to denatured globin or to one of a great many nitrogenous compounds, such as pyridine, nicotine, etc. These various haemochromogens gave spectra of the same type, but differing in detail. According to Keilin the rather complex spectrum of cytochrome consists of six absorption bands, three of which are so close together as apparently to fuse into a single band. This spectrum can be resolved into three separate spectra, each of two lines. Of the three two appear to be those of different haemochromogens. It is not my purpose here to discuss the third spectrum. The point is that over a great part of the animal and vegetable creations haematin is to be found in association with some nitrogenous compounds, on the same lines on which it unites with globin. Nor is this the whole story. Out of Keilin's research on cytochrome arose delicate methods of testing for haematin itself. The haemochromogen spectrum is seen in higher dilutions than that of any other blood pigment, and among haemochromogen spectra that of pyridine haemochromogen is pre-eminent for its visibility in low concentrations. Where haematin is suspected, therefore, it is only necessary to add pyridine in the presence of a reducing agent, and if haematin is present, even only in infinitesimal quantities, it will stand revealed.

Haematin has thus been discovered in many unexpected places, of which the growing points of certain vegetable tissues are not the least interesting. The interest lies largely in the fact that as the cell ages the haematin appears to be transformed into cytochrome; haematin is therefore the more primitive material; and if the vegetable cell had contained the necessary globin, there is no reason why haemoglobin should not be found widely distributed in the vegetable kingdom.

But to pass from the evolution of haemoglobin itself to that of its peculiar properties. The question naturally arises: Is the sigmoid oxygen dissociation curve which so resembles a titration curve in type an inherent property of haemoglobin, or is it something rather specialised?

The answer to this question has engaged the thoughts and energies of many interested in it for almost two decades.

The following considerations, however, appear to be worth some thought.

(1) In *Helix pomatia* there exists a form of haemocyanin in which the affinity of the pigment for oxygen is unaffected by the hydrogen-ion concentration of the medium in which it is dissolved. The general parallelism between the properties of oxyhaemoglobin and oxyhaemocyanin is so close as to encourage the belief in the similarity of the oxygen linkage in the two compounds. There is no evidence of any sigmoid inflection of the dissociation curve of this oxyhaemocyanin in *Helix pomatia* (see Fig. 50).

In the meantime one does not easily escape from the comment made by Dr and Mrs Stedman (1928): "In view of the history of the oxygen dissociation curve of haemoglobin there will necessarily be some hesitation in accepting the implications of the results of the foregoing experiments". The experiments in question are those illustrated in Fig. 50. The allusion is of course to the fact that a dissociation curve for oxyhaemoglobin which is quite devoid of double inflection has never with certainty been produced. In 1914 we seemed to be "nearly there". It looked as though we had only to obtain haemoglobin a little purer than hitherto in order to obtain a specimen which yielded a hyperbolic dissociation curve.

(2) Haemoglobin made by R. Hill (1933), by the addition of haematin to globin, appears to have no inflection in its dissociation curve, and the same is said to be true of the haemoglobin of the eel (Redfield, 1933) and indeed of ordinary haemoglobin on "ageing".

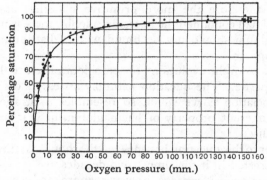

Fig. 50. Oxygen dissociation curve of blood of *Helix pomatia*, points range from pH 4 to pH 9.

So far we have considered haemoglobin as being simply a constituent of blood, but the amount of haemoglobin situated in muscle is in some animals quite considerable. It appears in respect of iron closely to resemble the haemoglobin of blood. In the lowest forms of life in which haemoglobin is found, its distribution is wider still: "In the annelid *Aphrodite aculeata* (Lankester) it is found in the nerve ganglia. The colour is most intense in the supraoesophageal ganglion which has as intense a colour as a drop of human blood.... An exactly similar observation has been made by Hubrecht who found haemoglobin in the red coloured cerebral matter of certain worms which possess no coloured blood corpuscles" (Schafer,

1898). What the function of haemoglobin in these lowly forms may be is quite obscure. Various workers have suggested that it holds a store of oxygen as against the times when the supply of that gas may be cut off.

In recent years the trend of opinion has been against this view, possibly because exaggerated claims were made for it. It always seemed to me suggestive, I cannot say more, that the lugworm, *Arenicola*, contained about as much haemoglobin in its body as could supply its consumption of oxygen from one tide till the next. Man carries in his blood less than five minutes' supply of oxygen, *Arenicola* five hours' supply—why?

In the mammals and birds, moreover, the only explanation which till recently could be given of the haemoglobin in muscle comes very near to the idea of a store. The occurrence of haemoglobin is characteristic of those muscles which undergo rhythmic contractions over long periods of time—such are the heart muscle of the larger mammals and the wing muscles of birds. Now it is of the nature of a muscle that when it contracts it should require more oxygen; on the other hand its supply is temporarily cut off because the pressure of the contracting fibres pinches the blood vessels. This has been beautifully shown by Anrep and his co-workers for both the heart (1927) and skeletal muscle (1934). The actual internal medium of the contracting muscle would suffer great oscillations in the content of dissolved oxygen. But by enriching that medium with a copious supply of haemoglobin the contracting elements can be supplied with a continuous supply of oxygen, even though this comes from a discontinuous source.

Yet when we have given what credit we can to the properties of haemoglobin there remains the fact that the "flat" part of the curve is only a third of the whole. If the ordinary utilisation of oxygen reduces the saturation of the haemoglobin from 97 to 65 per cent., that utilisation could be doubled, but not more than doubled, without greatly reducing the oxygen potential of the plasma.

In point of fact the oxygen requirements of the muscle during exercise are much more than double that during rest, and therefore additional machinery is necessary if the pressure of oxygen in the plasma of the venous blood is to remain in the region of 34 ± 7 mm.

The method on the larger scale of preserving the constancy of the oxygen content of the venous blood is the supply to the tissue of a more copious flow of blood, involving as it does the alteration in calibre of the vessels which irrigate the tissue: involving also consequential alteration in the rate of the heart, the quantity of blood in remote organs, and many other things. Into the detail of these alterations it is not my purpose here to enter. I wish to point out that we have taken the same jump as in the consideration of the higher regulation of respiration and of temperature, the jump from a purely chemical and palliative process to one of elaborate nervous complexity, which primarily involves the sympathetic system.

BLOOD SUGAR

The blood sugar in the lower vertebrates is much more variable than in man. In other words the internal "milieu", so far as glucose is concerned, exhibits much less "fixité".

Observations in fishes have been made by Hall and Gray (1929), who have kindly furnished me with the following data:

```
Goose fish      0 –10    mg. sugar per 100 c.c. blood
Scup            35·3–116·2        ,,        ,,
```

So much for cold-blooded animals; in ducks Seitz (1929), working in the North-Western University, Chicago, finds the normal range of blood sugar to be 100–160 mg. per 100 c.c. of blood. Chickens appear to be even more variable. Koppanyi, Ivy, Tatum and Jung (1926) give the normal maximum as 200–350 mg. glucose per 100 c.c. of blood. Holmes and Holmes (1927) give a few figures for normal mammals:

```
Cats (3)       123, 110 and 118 mg. per 100 c.c. blood
Rabbits (2)    160 and 150          ,,        ,,
```

The percentage of sugar in human blood is kept at a remarkably constant level. As in the normal man no appreciable quantity of sugar is secreted in the urine, it follows that approximately all the carbohydrate eaten is sooner or later oxidised. In the meantime it is stored in the liver and in the tissues. The constancy in the blood is then assured by:

(*A*) The abstraction of sugar from the blood if an excessive amount is thrust into it as after a carbohydrate meal. This excess is taken up by (1) the liver, and (2) the muscles.

(*B*) The contribution of sugar to the blood from (1) the liver, and (2) the muscles (indirectly *via* lactic acid), where this or that tissue oxidises the store and depletes the blood to make good that store.

It is not possible to make a final statement as to the mechanism by which either or both of these processes are regulated, but clearly if we trace back the regulation far enough we arrive at the central nervous system. Claude Bernard located the centre controlling the conversion of glycogen into sugar in the medulla, in the region of the *noeud vital*, but as with the respiratory centre there is undoubtedly an ultimate control at a higher level. This for rabbits has been placed by Donhoffer and Macleod (1932) in the pons, but several other reliable observers, Beattie, Brow and Long (1930), Bulatao and Cannon (1925), Mellanby (1919), Olmsted and Logan (1923), Keeton and Becht (1915), Karplus and Kreidl (1927), Houssay and Molinelli (1925) and Himwich and Keller (1930), all of whom have worked on either dogs or cats, find evidence of a regulating mechanism at a higher level than the pons, either in the hypothalamus or in the cerebral cortex itself or in both.

GENERAL DISCUSSION AND CONCLUSIONS

Now I must turn for a moment to the consideration of what to me is the most interesting aspect of Claude Bernard's famous statement: "The fixity of internal environment is the condition of a free life". Authors of great distinction have emphasised the importance of the statement, they have stressed the fixity of the environment, but there is a curious silence about the freedom of the life. What has the organism gained by constancy of temperature, constancy of hydrogen-ion concentration, constancy of water, constancy of sugar, constancy of oxygen, constancy of calcium and the rest? Is not the poikilothermic animal a very good animal? The bass or the perch with a blood on the acid side of normal, what fault is to be found with its muscular contraction or its heart beat? You may

take other organs of the body and subject them to a concentration
of oxygen far below that necessary for man as a whole and they will
function excellently. What then is this free life of which the fixity
of internal environment is the condition?

One approach to the investigation might be to consider what
happens to man if the higher or lower limits of the fixity are passed
over.

A brief statement of the results will be found in Cannon's
(1929) article on homeostasis. This I have taken as the base of

Table IV

Environment	Deficient	Excessive
Temperature	Inertia	Delirium
Oxygen	Unconsciousness	—
cH	Headache	Coma
Glucose	Nervousness	—
	Feeling of "goneness"	
	Hunger	
Water	(Weakness, Asher)	Headache
		Nausea
		Dizziness
		Asthenia
		Inco-ordination
Sodium	Fever	Reflex irritability
		Weakness
		Paresis
Calcium	Nervous twitchings	Apathy
	Convulsions	Drowsiness verging on coma
		General atonia

Table IV, in which the data are either taken directly from Cannon
or from the references which he gives, except in a few cases when
the information was deficient.

Glance down this list and tell me what these symptoms have in
common. Possibly my mind reacts to them rather more rapidly
than those of some because fate has forced a number of these
deficiencies and excesses on my own person. The recollections
which such experiences recall are quite clear cut.

What comes back when I recall the attempt to reduce my body
temperature? About the effects on the heart I have told you; they

were interesting but in no way arresting, but what comes back is the effect on my mind. In each of the two experiments which I performed there was a moment when my whole mental outlook altered. As I lay naked in the cold room at Woods Hole I had been shivering and my limbs had been flexed in a sort of effort to huddle up, and I had been very conscious of the cold. Then a moment came when I stretched out my legs; the sense of coldness passed away, it was succeeded by a beautiful feeling of warmth: the word "bask" most fitly described my condition; I was basking in the cold. What had taken place, I suppose, was that my central nervous system, or at least the subthalamic portion of it, had given up the fight, that the vaso-constriction had passed from my skin, and that the blood returning thither gave that sensation of warmth which one experiences when one goes out of a cold-storage room into the ordinary air. I suppose too, that had the experiment not ended at that point my temperature would have fallen rapidly and that I was on the verge of the condition of travellers when they go to sleep in extreme cold never again to awake. And I was conscious of other reversions of mental state; not only was there a physical extension of the limbs, but with it came a change in the general mental attitude. The natural apprehension lest some person alien to the experiment should enter the room and find me quite unclad disappeared—just as flexion was changed to extension, so the natural modesty was changed to—well I don't know what. Clearly one should be very cautious about taking these liberties with one's mind—and that is the point, the higher parts of the central nervous system were the first things to suffer.

So much for cold, and now for heat. What happens if the body temperature rises? I suppose there are few of my readers who have not at one time or other experienced this variation in environment —the result is delirium, before the heart has lost its efficiency or the respiration is more than quantitatively affected the coherence of the mind has gone. In the case of many fevers the effect of toxins must be eliminated before any conclusions can be drawn, but in heat stroke no such considerations come into play.

Heat stroke. Little systematic work appears to have been done with regard to the relative effect of heat in the mental processes of man and of animals. Rice and Steinhaus (1931) caused dogs to swim in water ranging from 15° to 40° C., and so produced a variation in the body temperature between the limits of 35° (95° F.) and 43° C. (109° F.) respectively. As regards the mental condition of these animals no explicit statement is given, but the following sentence perhaps throws a sidelight: "Complete fatigue was considered achieved...when his movements were insufficient to keep his head above water". Clearly, although the mentality may have deteriorated the dogs were conscious and sane (if I may use the term). The same would not have held of man, even had he been alive.

I have seen two dogs standing side by side one with a temperature of 88·5° F., the other with a temperature of 105·0° F. In neither case was the mental condition duller than that of many normal dogs.

I am indebted to Dr Frank Marsh, author of "Etiology of Heat Stroke and Sun Traumatism" (1930) for some information about the effects of high temperatures on rabbits. He writes: "The rectal temperature rose to between 109·5 and 112° F. in the course of three-quarters of an hour. During the period when the temperature was rising there were few symptoms....Over 107° F. the rabbit's face assumed a peculiar anxious expression, it was not unconscious but supremely indifferent, it took no notice of water when offered to it although it was obviously thirsty....Somewhere between 108° F. and 110° F. most of the rabbits lost consciousness".

From the paper named above (Marsh, 1930) I quote: "The experimental heat hyperpyrexia observed in rabbits very closely simulated the 'heat stroke' observed in man in this country (Iraq) and elsewhere.

"The nervous system of the rabbit resists heating to 109° F. for 1¼ hours. The temperature of collapse varied slightly in different rabbits, in eleven experiments it equalled or exceeded 110° F. Man usually collapses at 104° F. or 105° F."

Probably the literature would provide other and abundant information calculated to show within how narrow a range of temperature the mental processes of man are restricted, as compared with the more simple mental processes of the lower (but not much lower) animals.

So much for temperature.

Hydrogen-ion concentration is a quality of the blood which may be reduced with little difficulty; it is only necessary to breathe with sufficient violence for a sufficient length of time. Try to pant as violently as you can for three minutes and if you are not "fuzzy in the head" at the end of that time I shall be greatly surprised.

This question arises, however: Is the fuzziness the direct result on the cerebrum of the washing out of CO_2, or is it due to cerebral anaemia secondary to a general fall of blood pressure?

Increase of hydrogen-ion concentration. The simplest way of increasing the hydrogen-ion concentration of the blood is the inhalation of an atmosphere which contains an abnormally high percentage of carbonic acid. The highest percentage which I know to have been breathed experimentally was eleven, and that for a short time by Mr J. B. S. Haldane. Dr Margaria and I have been in an atmosphere of 10 per cent. CO_2 in air for perhaps 5 min., but I think that somewhere around 7 or 8 per cent. is as much as can be withstood for any extended period—and with that I think Mr J. B. S. Haldane would agree. Margaria and I spent about 20 min. in 7·2 per cent. CO_2, and were quite ready to come out at the end. Our symptoms were rather different, but in both cases they were connected with the highest parts of the central nervous system. Margaria suffered from headache for the rest of the day, and Haldane was affected in the same way: my own symptoms were no less definite, but were those of mental fatigue—inability to concentrate on or even listen to conversation without an effort; the tendency to take up a newspaper, read a few lines of one paragraph, preferably something quite unimportant, then a few lines of another without finishing anything, and so forth. This was associated for a few hours with a feeling which was not exactly hunger and not exactly nausea, but a sort of mixture of the two: the day following—a Saturday—my wife, who had not known of the experiment, said, "I cannot understand why you seem so tired: I am going to keep you in bed for the weekend"; the mental symptoms lasted about 2 days. I took some samples of my own alveolar air in the chamber at intervals. Analysis proved that there had been errors of manipulation in the last two samples. Now the interesting point is not that these errors occurred, though that is quite significant, but that I could have gone into a court of law and sworn that one at least of the two was correctly taken. On the occasion on which we were in 10 per cent. CO_2 I was, when I came out, retaining my grip of things only with an effort. Margaria and I agreed on two things, firstly that we did not want to repeat this experiment unless there was some good reason for doing so, and secondly that our reluctance was due to our unwillingness to expose the higher parts of our brain to the influence of so much carbonic acid.

Oxygen want. The symptoms of oxygen want are too well known to need much recital. Such as they are I have experienced most of them.

(1) *Acute.* In its most acute form want of oxygen first affects the reasoning portions of the brain; if persisted in unconsciousness follows. My mind in this connection reverts to an experiment which I was doing in company with Alfred Redfield. I was on a bicycle ergometer, and certain manipulations of taps were required of me. I was breathing nitrogen, or at least some mixture poor in oxygen. When the time came to turn the taps I made no movement. Redfield at once realised that my mind had become too confused to know what to do, and directed me to carry out each process. Under his direction there was no difficulty in observing the ritual.

Many other instances might be given of similar lapses. One of the most striking is that in which the late Sir Clement le Neve Forster, partially overcome with carbon monoxide in a mine, sat committing to paper lengthy farewells to his family when he knew that he had only to get up and walk about 20 yards to be in a place of safety.

(2) *Chronic.* The more chronic forms of oxygen want have been the object of abundant study. It is only necessary to say that those which beset the dweller at high altitudes are symptoms of the brain. They may be referred to other parts of the body, palpitation of the heart, breathlessness, vomiting, etc. A moment's consideration shows that breathlessness is not an affection of the chest, but of the

nervous mechanism which operates the diaphragm and the intercostal muscles.

Oxygen want, as has been shown by Greene and Gilbert (1921),
never accelerates the perfused heart. Its action is in the opposite
direction, and in order to affect the perfused heart a degree of
anoxaemia is required which the brain would never tolerate. The
palpitation induced by exercise at high altitudes is nervous.

Apart, however, from these features there are others affecting
higher parts of the brain. I remember one of our party saying after
we had left Oroya: "Most of us suffered from mountain sickness,
we were all interested in its relation to oxygen want, we all knew
that there was an abundance of oxygen cylinders at hand, yet none
of us thought of trying to see whether oxygen inhalation would rid
us of our symptoms—that was characteristic".

Oxygen excess. It seems doubtful whether the effects of exposure
to excessive concentration of oxygen can rightly be discussed in an
article on the evolution of the constancy of the internal environment. For my reticence there are two reasons. The first is that as
the organism has never in its evolution had to meet an excessive
pressure of oxygen such a contingency can have no place in
evolution. The second reason is that oxygen being something which
is inhaled is, in its relation to the lung, part of the external environment, and the fact that it produces pneumonia in rats and mice
might be deemed irrelevant.

Yet with these reservations it may not be amiss to record the little
that is known concerning the effects on the organism of a too high
concentration of oxygen in the blood. They were first discovered
by Paul Bert (1878), and have been confirmed so far as some of the
meaner creatures are concerned by Lorrain Smith (1899). With
regard to man there is no information on the subject.

Two larks were placed in a chamber, and the oxygen raised to a
tension of 301·4 per cent. of an atmosphere. They at once became
excited, and moved rapidly about the chamber. After 13 min.
exposure to this tension, they were simultaneously thrown into
violent convulsions. These recurred at short intervals. They began
to subside in about an hour. After 2 hr. 7 min. the chamber was
opened. One of the birds remained in an unconscious condition

with occasional epileptiform convulsions for about 1 hour after, when it died. The other survived, and was very active and restless for a while, but became very sluggish later. When it was fed by the hand, however, it shook off its drowsiness for a short time, and again assumed its normal activity. It survived in this condition for several days.

Normality then takes place between limiting concentrations of a number of materials or between limiting physical conditions. If the limits are transgressed something happens to impair the efficiency of the organism. Look down the list of disabilities produced by alterations in the internal *milieu* (p. 80) and you will see almost no reference to the grosser bodily functions, nothing about muscular contraction as such, nothing about the heart as such, nothing about the kidney, the liver or the pancreas. In almost every case the blow is to the nervous system; we can go further, in almost every case it is to the central system. In our discussion of temperature regulation we drew attention to the similarity between the effects of cooling the lightly anaesthetised cat and those designated by the Harvard school as " false rage "—effects which Bard has so successfully located in the hypothalamus; we may now go a stage further and compare one of the most recent pictures of that region (Fig. 51) with the table on p. 80. The comparison will show how little of the contents of this table are omitted from the picture—that little appears likely to be situated above rather than below the region in question. The fixity of internal environment, then, is controlled by the upper part of the central nervous system, and it is as a general rule the upper part of the central nervous system which suffers if the environment alters beyond physiological limits. The fixity of the internal environment is in short the condition of mental activity.

The chemical and physical processes associated with the working of the mind are of so delicate a character that beside them the changes measured by the thermometer or the hydrogen electrode must be catastrophic—overwhelming. Processes (probably rhythmic) of such delicacy must surely require a medium of great constancy in which to attain to ordered development. How often have I watched the ripples on the surface of a still lake made by a passing boat, noted their regularity and admired the patterns formed when two such ripple-systems meet—but the lake must be perfectly calm, just as the atmosphere must be free from atmospherics if you are to find beauty in the subtle passages of the symphony. To look for a high intellectual development in a *milieu* whose properties have not become stabilised is to seek music

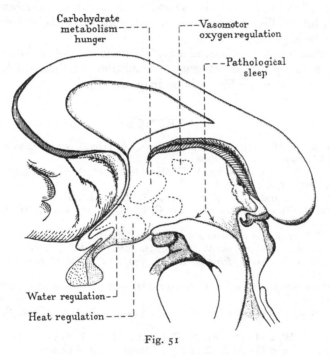

Fig. 51

amongst the crashings of a rudimentary wireless or the ripple-patterns on the surface of the stormy Atlantic. Just as an army, after the tedious ascent through a mountain defile, may arrive at and deploy over the open plateau of the summit, so it is not unnatural that the slow and laboured course of evolution should be broken at intervals by sudden and rapid developments. Such would appear to be the sequence before us; by degrees throughout the ages the constancy of the internal *milieu* became controlled with ever-increasing nicety till ultimately this control was perfected up to the point at which man's faculties could develop and he could attain to an understanding in terms of abstract knowledge of the world around him.

Each century, and now each decade, add emphasis to the antithesis between the complete insignificance of man when considered as part of the material universe and the astounding ascendency to which his intellect has attained in comprehending the universe in which he is placed.

Of that intellectual ascendency, "la fixité du milieu intérieur" appears to be the, or at least a, condition; of that intellectual ascendency—"la vie libre" is no inapt description.

REFERENCES

ANREP, G. V., BLALOCK, A. and SAMAAN, A. (1934). *Proc. Roy. Soc.* B, **114**, 223.

ANREP, G. V., CRUICKSHANK, E. W. H., DOWNING, A. C. and RAU, A. S. (1927). *Heart*, **14**, 111.

ANSON, M. L. and MIRSKY, A. E. (1925). *J. Physiol.* **60**, 50.

BEATTIE, J., BROW, G. R. and LONG, C. N. H. (1930). *Proc. Roy. Soc.* B, **106**, 253.

BERT, P. (1878). *La Pression Barométrique.* Paris.

BULATAO, E. and CANNON, W. B. (1925). *Amer. J. Physiol.* **72**, 295.

CANNON, W. B. (1929). *Physiol. Rev.* **9**, 399, and *The Wisdom of the Body* (1932), London.

CHRISTIANSEN, J., DOUGLAS, C. G. and HALDANE, J. S. (1914). *J. Physiol.* **48**, 244.

DONHOFFER, C. and MACLEOD, J. J. R. (1932). *Proc. Roy. Soc.* B, **110**, 125.

GREENE, C. W. and GILBERT, N. C. (1921). *Amer. J. Physiol.* **56**, 475.

HALL, F. G. and GRAY, I. E. (1929). Communicated privately.

HILL, R. (1933). Verbal communication.

HIMWICH, H. E. and KELLER, A. D. (1930). *Amer. J. Physiol.* **93**, 658.

HOLMES, B. E. and HOLMES, E. G. (1927). *Biochem. J.* **21**, 412.

HOUSSAY, B. A. and MOLINELLI, A. E. (1925). *C.R. Soc. Biol.* **93**, 1454.

KARPLUS, J. P. and KREIDL, A. (1927). *Pflügers Arch.* **215**, 667.

KEETON, R. W. and BECHT, F. C. (1919). *Amer. J. Physiol.* **49**, 248.

KEILIN, D. (1925). *Proc. Roy. Soc.* B, **98**, 312.

—— (1926). *Ibid.* **100**, 129.

KOPPANYI, T., IVY, A. C., TATUM, A. L. and JUNG, F. T. (1926). *Amer. J. Physiol.* **76**, 212.

LEVY, C. L. (1889). *Zeitschr. Physiol. Chem.* **13**, 309.
MACMUNN, C. A. (1886). *Phil. Trans. Roy. Soc.* **177**, 267.
—— (1889). *Zeitschr. Physiol. Chem.* **13**, 497.
MARSH, F. (1930). *Trans. Roy. Soc. Trop. Med. and Hygiene,* **24**, 257.
MELLANBY, J. (1919). *J. Physiol.* **53**, 1.
OLMSTED, J. M. D. and LOGAN, H. D. (1923). *Amer. J. Physiol.* **66**, 437.
REDFIELD, A. C. (1933). *Quart. Rev. Biol.* **8**, 31.
RICE, H. A. and STEINHAUS, A. H. (1931). *Amer. J. Physiol.* **96**, 529.
SCHAFER, E. A. S. (1898). *Text-book of Physiology,* **1**, 187. Edinburgh.
SEITZ, I. J. (1929). Communicated to me by Dr Ivy.
SMITH, J. LORRAIN (1899). *J. Physiol.* **24**, 19.
STEDMAN, E. and E. (1928). *Biochem. J.* **22**, 897.

CHAPTER IV

STORES I (CERTAIN PROXIMATE PRINCIPLES)

The combination of a constant internal environment and an intermittent source of supply necessitates a storage of materials, on which the environment can draw.

Were these stores all evident the present chapter would be short and simple, for the mere enumeration of them would suffice. Some materials however are difficult to trace once they get into the body, a fact which I have heard expressed by Laurence J. Henderson in conversation, in some such phrase as "the body seems to contain what may be likened to marshes or swamps into which substances may disappear and be lost to view".

In considering these stores the most evident may be first considered. When I commenced to think of the subject, a portrait came into my mind of Sir Michael Foster, by Collier I think; behind him was a blackboard on which was written:

Carbohydrates,
Fats,
Proteins,
Salts,
Water.

At least that is my memory, and if I am at fault in the order of the substances enumerated, the order is a very suitable one for the present purpose.

CARBOHYDRATES

The storage of carbohydrate is in some ways typical of its kind, the facts with regard to it are so well known as to need little comment. Claude Bernard set it forth that carbohydrate was ingested intermittently, that it was turned into sugar in the alimentary canal, that it was absorbed as such by the blood of the portal system, that the excess of sugar in the portal system over that in the general circulation was laid up in the liver as glycogen, that as the muscles used up the sugar from the blood the glycogen was reconverted into sugar in order to make up the deficit and so kept the internal environment constant. The clear vision of these facts was no doubt due to the fact that all the substances involved gave distinctive chemical tests. Had not glycogen given the brown test with iodine and had not sugar reduced Fehling's solution, the carbohydrate might well have disappeared into one of Laurence Henderson's "marshes", as indeed it did before Claude Bernard recognised the true nature of glycogen. As it became possible to test both qualitatively and quantitatively for carbohydrate, we could follow the sugar in the portal blood and in the general circulation and also the glycogen in the liver and the other tissues in which it is stored.

The storage of sugar as glycogen appears to be a property of quite primitive protoplasm, it takes place in the Amoeba and the white blood corpuscle. So far therefore as the tissues of the mammalian body are concerned their relative power of storing glycogen is merely a question of the degree of differentiation which has taken place. In this respect clearly the liver stands out most prominently, being able to accumulate a much higher percentage than any other organ. It used to be said that the liver contained half of the glycogen in the body. The following data, however, given by Lusk (1928) puts the matter in a different light. "For 100 grams of liver glycogen there occurred in the rest of the body the following amounts":

Dog I	...	...	398 gm.
II	...	...	279
III	...	...	87
IV	...	...	76
V	...	...	159
VI	...	...	355
VII	...	...	105

The body in this as in so many other respects does not put all its "eggs in one basket" and that is perhaps characteristic of its attitude towards stores in general.

Hitherto I have spoken designedly of the hexoses under the general term sugar. We must not, however, pass over the fact that the body can lay hold of quite a number of nearly allied substances, turn them all into glycogen—so far as is known not into specific glycogens in the liver, but into one and the same substance—and preserve them for reconversion into a single sugar. "The writer personally prepared fructose from inulin in 1889, which when given to a fasting rabbit caused the formation in its liver of large quantities of glycogen, the anhydride of glucose. To a lesser extent the same fate *may* befall ingested galactose" (Lusk). So much for the three hexoses to which maltose, sucrose and lactose are reduced. There seems to be more doubt about the conversion of galactose than of the other two hexoses and possibly the word "may" which I have italicised is used quite designedly by Lusk. It would be satisfactory at this point to state the problem in a quantitative way. What is the concentration in the portal blood below which the liver does not take glycogen from the blood and above which it does? What is the concentration of sugar in the portal blood below which the liver contributes sugar to the blood and above which it does not? Are these concentrations the same? Does the rate at which the liver takes up glycogen vary in any simple way with the concentration of (a) sugar in the blood, or (b) glycogen in the liver? Does the rate at which the liver turns glycogen into sugar bear any simple relation to the concentrations (a) of glycogen in the liver, or (b) of sugar already in the blood?

In another respect the picture is incomplete!

(1) Granted that the sugar in the portal vein is too high, how does the liver appreciate that fact and "get to work", turning sugar into glycogen?

(2) Granted that the sugar in the general circulation is too low, how does the liver cell appreciate the fact and turn glycogen back into sugar?

(3) It is possible to approach the subject from the chemical aspect and ask whether some balanced action between the liver cell glycogen and the liver cell sugar is responsible for the existing state, but I gather there is at present no answer forthcoming along this line.

I have commenced the consideration of "stores" with that of glycogen because, regarding the "store" as a definite feature in the architecture of function, as it might be an arch, this particular store is rather completely exposed to view, we can see it not only as an arch but as an arch of a particular type—as it might be Gothic or Norman. It is not only an architectural feature but a feature with a well-defined type of architecture, and we ask ourselves whether other storage processes conform to this type or no.

The features are, that slightly differing carbohydrates reach the store as relatively small molecules in solution—molecules capable of easy and simple reactions, that they are stored as molecules of something which is of high molecular weight and of approximately colloidal characteristics—it is removed from the store in approximately, but not necessarily quite the form in which it arrived, as single, small molecules of dextrose, and is thus transported to the seat of action; there it is or is not—according to opinion— reconverted into glycogen before use. So far as is known the conversion processes are such that enzyme action plays a large part in them. To what extent are these general features discoverable in the storage of other materials? Have we arches of different architectural styles and if so what can we say about them?

FATS

Turning to the storage of fats. Can all the fat of the body be regarded as "stored", that is, as materials which can be drawn upon for the provision of energy by the tissues?

The question has been the subject of comprehensive studies by two French physiologists, Terroine (1914, 1919) and André Mayer (1907, 1913, 1914), for whom I have a special respect; I was very closely associated with them in another field of physiology (or pharmacology) during the Great War. Their general thesis is as follows:

(1) If animals be starved a portion but only a portion of the fat disappears and is used for the production of energy. That portion is called the *élément variable* as opposed to the *élément constant* which remains. Only the *élément variable* can be regarded as in "store".

(2) *The élément constant* is regarded as part of the essential architecture of the cell. If I may quote Leathes and Raper (1925):

> André Mayer and his fellow-workers then have done much to substantiate the conception of cell protoplasm that Hoppe Seyler, and Hermann, indicated many years ago. Protoplasm is not merely an old word for a solution of proteins; nor are the phenomena of life manifested merely in a colloidal solution of proteins in which certain electrolytes and certain reserve foodstuffs play merely subsidiary parts. Protoplasm is a complex equilibrated system in which, side by side with colloidal solutions of proteins, fatty components together with cholesterol, though quantitatively less in amount, play a rôle which it is impossible to say is less essential than theirs.

The fat so built into the cell is no part of the body store, and it seems to have chemical properties of an interesting character which centre around the element phosphorus. For the most part the ratio of phosphorus to fatty acid is more or less constant, being of the order of 1 : 18 by weight—which suggests one atom of phosphorus to two molecules, or rather radicles, of fatty acid, and that is the condition of affairs in lecithin.

$$
\begin{array}{l}
\overset{\displaystyle O}{\underset{\displaystyle \|}{}} \\
H_2=C-O-C-R_1 \\
\overset{\displaystyle O}{\underset{\displaystyle \|}{}} \\
H-C-O-C-R_2 \\
\overset{\displaystyle O}{\underset{\displaystyle \|}{}} \\
H_2=C-O-POCH_2 \\
OH CH_2 \\
N(CH_3)_3 \\
OH
\end{array}
$$

(Lecithin)

Unless therefore phosphorus is present in quite other combinations, there seems to be little room for the glycerides of acids which do not contain phosphorus among the fats which form the framework of the tissues. The glycerides free from phosphorus are to be found in the stored fat—in what we ordinarily recognise as "fat", in the subcutaneous tissue, the omentum and elsewhere.

Having now drawn a line between stored fat, the *élément variable* and intrinsic fat, the *élément constant*, we are in a better, though still not a good, position to ask whether in its general plan the storage of fat has much in common with that of carbohydrate.

Carbohydrate is largely ingested in the form of the insoluble material starch; fat enters the alimentary canal entirely in a form insoluble in water.

Both fat and carbohydrate are reduced to a soluble form in the alimentary canal and I imagine there is now a general concensus of opinion that fat like carbohydrate passes through the lining surface of the intestine in a soluble form. But does the parallelism hold any further?

(1) Carbohydrate is found in blood in a soluble form, can the same be said of fat?

(2) Carbohydrate is stored in an insoluble form, can that also be said of fat?

(3) Carbohydrate is released from store in solution, does that also hold for fat?

In the blood two substances are present which are not simply compounds of glycerine and fatty acid; one is the cholesterol compound and is a "wax", that is a compound of fatty acid with a monohydric alcohol; the other is lecithin, a compound of fatty acid, phosphoric acid, glycerine and a base choline. Table V, given by Macleod (1930), shows the distribution of fatty acid in normal blood:

Table V

	Total fatty acids %	Lecithin %	Lecithin in terms of fatty acid %	Cholesterol %	Glycerides %	Total lipoids %
Whole blood	0·37	0·30	0·20	0·22	0·10	
Plasma	0·39	0·21	0·14	0·23	0·10	0·068
Corpuscles	0·34	0·42	0·28	0·20	0·0	

From it we learn that the whole quantity of fat in the plasma is perhaps 4 gm. per litre, or about 10 gm. in the whole plasma of the body and of that only about one-quarter consists of the glycerine compounds of the fatty acid. The fat, it would seem, gets transformed for purposes of transport.

By comparison of the fatty acid, lecithin and cholesterol contents of the blood *during fat digestion* it has been found that there is a steady but very variable increase of fatty acid, accompanied by an increase in the lecithin, which varies from 10 to 35 per cent., but does not run strictly parallel with the fatty acid increase. There may also be an increase in the cholesterol. It looks as if fat were absorbed into the corpuscles where it is transformed into lecithin which is then returned to the plasma, cholesterol also appearing when the lecithin reaches a certain concentration. The fat then in these forms which are relatively (to stearin) soluble in some at all events of the constituents of the blood are transferred to the fat depôts and there deposited, but not without reconversion in the fat cells into the glycerine esters of the fatty acids.

The picture given above by Macleod is based largely on the observations of Bloor. Bloor (1922) believes that the major part of the fat which is about to enter the fat depôt reaches it as lecithin; he points out, for instance, that Meigs and his co-workers can account for the entire amount of fat secreted in the milk on the basis of the loss in lipoid phosphorus by the plasma in its transit through the mammary gland. The armoury of the cell for dealing with lecithin is, moreover, more complete than that for dealing with fat. At some stage the fat is stored as the characteristic fat of the animal, this probably takes place in the cell of the depôt. In order that it may occur, the base must be split from the fatty acid—an analysis carried out by a ferment in the cell. But whilst lipase is not an enzyme of common occurrence outside the alimentary canal and the pancreas, the ferments capable of splitting lecithin are of wide distribution and are found in such places as the body stores fat. The analogy between carbohydrate storage and fat storage, on Bloor's theory, fairly complete though not absolutely so, may be summarised as follows:

The fat reaches the alimentary canal in insoluble form; carbohydrate does so to a considerable extent. Each becomes dissolved as the result of hydrolysis and passes into the wall in solution. In the intestinal wall the fat, unlike the carbohydrate, becomes re-converted into something like the original insoluble material. It finds its way into the blood mostly if not entirely through the thoracic duct. In the blood it is largely reconverted into a soluble form and notably a form soluble in the corpuscles, so that much of the fat like all the carbohydrate reaches the depôts in a form in which it can pass through the cell wall. Inside the cell, lecithin, at all events, is attacked by a ferment, the esterase, and is converted into the insoluble body fat proper to the animal.

The answers to the first two of the three questions asked on p. 95 would on that theory be definitely in the affirmative. Such is Bloor's picture of the transport of fat to the depôts, but it must be stated quite definitely that many of the most reliable writers do not accept it or at best accept it only partially.

In the first place, it gives the impression that all the fat in the blood is in solution. If the term solution be used in its strictest sense, that cannot be so. It is notorious that plasma after a meal of fat is milky, which means that fat is present in aggregates large enough to scatter light. To quote Leathes and Raper: "On micro-scopic examination of human blood Neumann (1907) found that ultramicroscopical particles were seen two hours after a meal con-taining fat, but not after a meal free from fat; nor were such particles to be found if the sample of blood were taken before breakfast".

Granting then that the fat found in the blood after a meal is at least infra-microscopic, one naturally enquires what evidence there may be, independent of Bloor's work, of

(1) The increase of total fatty acids in blood;

(2) The increase of lecithin;

(3) The increase of cholesterol.

So far as (1) and (3) above are concerned, Terroine has supplied such evidence. According to him not only are the total fatty acids increased but also the cholesterol; moreover, he regards the ratio of the latter to the former as being approximately constant in the same animal from time to time, this ratio he terms the "lipaemic index".

Bloor on the other hand believes the principal increase, at all events as late as the fourth to the seventh hours, is not in the cholesterol but in the lecithin.

If I have followed correctly the very complicated question of the transport of fat, the issue is something of this kind.

Most authorities would agree that there were four possible ways in which fat might be transported from the thoracic duct to the fat depôts, thus:

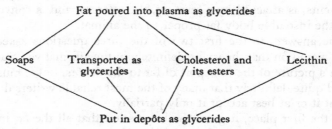

But authorities would emphatically disagree as to the relative quantity of fat, out of 100 gm. transported, which goes along each of the four paths. In other words, we do not know the kinetics of the process of fat transference.

Supposing, for instance, one accepts the view of Terroine that after a meal of fat the glycerides and the cholesterol in the plasma are increased in equal proportions, that is quite a different statement from saying that fat from the intestines is transformed at equal rates into glycerides and into cholesterol and its esters, for we do not know that these disappear from the blood at equal rates. Indeed, concerning their disappearance into the fat depôts—as Raper has pointed out to me—we know practically nothing. Something can be said with regard to the rate of deposition of glycogen in the liver, or at least there is this basis for a statement on the subject, namely, that the increase in percentage of glycogen in the liver is a measurable quantity, but when fat is deposited in the omentum there is no measurable increase in the percentage of fat present, it remains at about 95 per cent. Presumably more connective cells become fat cells.

Moreover, the situations in which fat is stored are very various, the main depôts are no doubt the subcutaneous tissue and the omentum, but on rather different lines fat, like carbohydrate, is to be found in the liver often in large quantities. What the length of its stay there, or precisely how far its presence can be described as one of storage, is perhaps beyond the scope of the present review. It is one on which opinions differ.

Again, there is this great distinction between the storage of fat in the depôts and the storage of carbohydrate in the liver, the liver-glycogen, so far as we know, comes from carbohydrate in the blood. Can we say that the radicals of fatty acid laid down in the depôts are all conveyed thither as fatty acid in the circulating fluid? Much of the fat which is found in the depôts no doubt had its origin not in fat taken but in carbohydrate—where does the change take place? Is the fat cell capable of making fat from carbohydrate? If fat can be laid down in the depôts as the result of the series of events which Bloor has described, and if we see in that series one closely parallel with the events which lead to the deposition of glycogen—if the arch resembles architecturally the carbohydrate arch, we must still admit that it is but a single arch in the very complicated and largely invisible structure of fat deposition. But if we know little about the carriage of fat from the intestine and its deposition in the fat cell of the depôt, we know even less about its final journey from the depôt to the place where it ceases to be fat, wherever that place may be. I suppose that it is not converted into carbohydrate in the cell of the fat depôt, but I do not know.

We know at least that the quantity of fat in the blood increases between the time that the fat is absorbed in the intestine and that of its storage in the depôt. We do not know for certain whether the fat in the blood is increased, decreased or changes in any definite direction between the moment that fat leaves the store and is used as energy. Does it leave the store because the tissue calls it, reducing the fat in the blood, or does it leave the store because it is thrust out? We know that the fat of the depôts can be used to provide energy of muscular contraction. Whether it is turned into carbohydrate first is a matter of opinion, and among those who regard this conversion as a necessary step, I imagine there is no unanimity as to where this step is taken.

The edifice we are endeavouring to study is so obscured as

to make its architectural type incapable of description with any certainty. We can at most say this, that in so far as Bloor's views form even a part of the truth, one arch bears a general resemblance to the other.

CALCIUM

In spite of the great volume of work which has been devoted to the subject of calcium metabolism within the last decade, knowledge of the processes involved in the storage of calcium is much less complete than it is in the case of carbohydrate.

It will be best to consider first the actual store. As in the carbohydrate and fat, the distribution of calcium in the body is not confined to a single place. Its distribution, as has been pointed out by Stewart and Percival (1928), is well-nigh universal, nevertheless, as in the case of those two substances, there is, *par excellence*, a store. As the liver is to glycogen and as the fat depôts are to fat, so the cancellous tissue of the bone is to calcium. This has been demonstrated in a series of researches by Aub and his colleagues Baur, Allbright and Hunter (for literature, see Hunter, 1931). The calcium of the bones ebbs and flows according to the diet of the individual. On a poor diet the bone yields calcium to the blood, as also when an excessive strain is put upon the calcium metabolism of the body. The history of Aub's discovery is not without interest. Our general conception of the march of knowledge is that fundamentals are first discovered as part of the progress of abstract science and that their application comes later. The soundness of that conception is unassailable. It would be a tragic day for learning if she ever had to justify investigation on the ground of immediate utility. But not a few cases can be cited in which the urgency of some immediate problem has led to the discovery of knowledge of a fundamental character. So it seems to be in the present case. The demands of industrial physiology compelled the investigation of lead poisoning. Lead is laid down in the cancellous tissue of the bones, from which it can be dislodged by parathyroid administration, with corresponding reduction in the cancellous tissue. If, however, the animals used are given relief from the parathyroid administration for a time and then it is reinstituted, calcium only is released from the bones and such lead as had been left in them is for a time at all events retained. It would appear that during the pause between the two periods of parathyroid administration a fresh layer of calcium phosphate has overlaid the now deeper lead-bearing bone—has become stored in fact—and not till the cancellous tissue has yielded up this recent store is further lead eliminated.

I have spoken of the cancellous tissue yielding its store when the animal is subjected to a low calcium diet, or to conditions which compel excessive calcium excretion. There is, however, one obvious contingency which must not be overlooked, namely pregnancy. It may reasonably be supposed that the demand of the foetus for lime might drain the blood of the mother, which drain in turn would compel the stores to disgorge their supply. Stewart and Percival state, with regard to the drain on the blood:

The best attested alteration in the serum calcium occurs during pregnancy and is doubtless due to the drain on the maternal tissues to meet the growing demands of the foetus, a demand which increases from 0·006 gm. per day during the first four months of gestation to over 0·6 gm. per day at term and may average 0·1 gm. over the whole period. Corresponding to this rapidly increasing foetal requirement the calcium in the maternal serum tends to fall, as Mazocco and Moron have shown in the following series of cases:

	Mg. Ca per 100 c.c. serum
10 healthy non-pregnant women	9·19
17 healthy puerperal women	8·79
29 pregnant women	8·77

The question arises, is this diminution in the calcium content of the blood associated with a decrease in the amount of calcium stored in the mother? On this subject Vignes and Croisset state "that in guinea pigs there is a progressive decalcification throughout the whole period of gestation".

This new view of bone as being not only a skeletal support but also as being a calcium store for the purpose of maintaining the calcium potential of the blood, as the liver maintains the sugar potential of the blood, is in itself sufficiently exciting, but from quite another quarter has come work which interdigitates into that of Aub and his colleagues in a very elegant way.

The work to which I allude concerns the process by which the actual laying down of the calcium in bone takes place; it was carried out by Robison and his colleagues at the Lister Institute and has reached its final fruition in collaboration with Miss Honor Fell at the Strangeways Institute. For some years the group, their head being of course Sir Charles Martin, had been keenly interested in the subject of rickets, with which they came in contact in a very practical way shortly after the War. They were closely associated with the amelioration of conditions in Vienna, where rickets was a prominent feature. The most reasonable preface to the investigation of why calcium was not laid down in the bones in rickets clearly was the mechanism of its deposition under normal circumstances. The general picture put forth by Robison and his colleagues (1923) is as follows:

When the fermentation of sugar takes place in the presence of phosphoric acid two acids are formed, hexose-diphosphoric acid and hexose-monophosphoric acid. The latter is the body to which he attaches importance in the present connection. This body, or rather its calcium salt, he regards as being present in small quantities in plasma and stored in relatively large quantities in the corpuscles—another store mark you.

The next point is that within the cartilage of the developing bone there is present a ferment which he calls phosphatase, which has the property of hydrolysing calcium hexose-monophosphate and producing as the result calcium phosphate in quantities greater than can ever be held in solution. This is deposited in the bone. The developing bone appears to contain ten times as much phosphatase as the rest of the bone.

Perhaps the most elegant turn taken by Robison's experiments was when, in collaboration with Miss Fell (1930), tissue cultural methods were applied to them. It is possible to observe the processes which lead up to ossification in explanted tissues.

It was found that the isolated femur of the 5–6-day embryonic fowl continued both its anatomical and histological development when explanted *in vitro*, underwent periosteal ossification and also synthesised a phosphatase which normally occurs in considerable quantities *in vivo* in bone and ossifying cartilage. Explants of the 3-day mesenchymatous tissue on the other hand gave rise to cartilage of a different type which neither ossified nor synthesised phosphatase.

The above experiments were extended to the bone-forming structures of the lower jaw for the purpose of observing whether there too the processes of calcification and formation of phosphatase went hand in hand. The comparison was between the non-ossifying cartilaginous distal portion of Meckel's rod and ossifying cartilage such as the palato-quadrate. Here again it was found that the normal processes could be carried out *in vitro*.

The distal part of Meckel's rod was removed from a number of 6-day embryos and explanted *in vitro*. Similar cultures were made either of the palato-quadrates or femora of the same embryos. As in the normal embryo the distal part of Meckel's cartilage when cultivated *in vitro* neither ossified nor formed a region of hypertrophic chondroblasts, whilst in the explanted palato-quadrates and femur an area of hypertrophic cells appeared which was usually associated with a layer of osteoid tissue. During cultivation the Meckel's cartilage synthesised no phosphatase, while the explanted palato-quadrates and femora developed high phosphatase activity.

The criticism has been brought against these experiments that there is on Robison's own showing one other place in the body where this ferment occurs in considerable quantities, namely, the kidney. The implication is I suppose that the kidney should calcify. But surely the kidney is a very special case, because in virtue of its very function things do not accumulate there, the condensation of phosphate is a primary function of the kidney, but the phosphate is carried away. May it not even be that the process of concentration and the phosphatase are not entirely unassociated?

Putting together the work of Aub and that of Robison, we arrive at a picture of calcium storage which is very much like that of carbohydrate storage. Calcium is carried in the blood in the form of a soluble compound of hexose-monophosphoric acid and in that form reaches the store, the cancellous tissue of the bones, where it is deposited as an insoluble material; a diet rich in calcium causes more rapid storage, whilst one poor in calcium leads to the withdrawal of the lime from the store; finally, the storage is wrought by the intervention of an enzyme phosphatase. The architectural features seem to be the same, but the accent in the picture is on the enzyme.

In the case of calcium storage the principal evidence for enzyme action is in the storing phase; in the case of glycogen it is in the

discharging phase. In the past, of course, the action of a liver ferment which turns glycogen into sugar has been hotly contested on
the ground that Claude Bernard's classical demonstration was a
post-mortem phenomenon, but I imagine that now there is a
general acceptance of Claude Bernard's thesis—namely, that the
everyday production of sugar in the liver is the result of enzyme
action.

It may be that the enzyme in each case is reversible in its action
and that according to the concentrations of sugar, calcium or
hexose-monophosphoric acid with which it is presented, it either
lays down the stored material or gives it out: but at the back of this
picture there is something else, namely, in the case of the liver,
insulin, in the case of the bone, parathormone. It may be, probably
it is, only a coincidence that each of these bodies acts on the phase
of storage which is opposite to that on which the evidence of enzyme
action is principally based, insulin facilitating the *storage* of glycogen
and parathormone the *discharge* of calcium. But each introduces
a factor extraneous to the storage system itself, and in the case of
insulin, as we saw in Chapter III, brings the whole process of
storage into the sphere of the nervous system.

WATER

The mechanism of the storage of water is particularly elusive.
The reason is not far to seek, water is everywhere in the body and
it is only by quantitative measurements that we can tell what has
become of any excess which the body may preserve against some
hour of special need.

What happens then to the water which is drunk? It is certainly
not immediately excreted.

Is it stored in the blood? Can it all be found there? If not, where
does it go? Can any of it be found in the blood, and if so, how
much? Such are questions which demand an answer, and at the
outset let me say that the answer probably depends upon circumstances, such, for instance, as the quantity of water which has been
taken and the time which has elapsed since the previous draught.

The question of what happened to water directly put into the
blood was investigated in some detail by several workers, who pre-

sent different pictures. Let us take first that of Engels (1904), who injected 0·6 per cent. (in one experiment 0·9 per cent.) saline into the jugular vein.

On the mean of seven experiments 1159 gm. of water were given per dog; of this on the average 352 gm. were excreted so that 807 gm. per dog were retained.

Haemoglobin estimation indicated that 31 gm. were retained in the blood:

$$\begin{array}{ll} \text{Retained in body} & \text{807 gm.} \\ \text{,,} \quad \text{in blood} & 31 \\ \text{,,} \quad \text{in tissues} & 776 \end{array}$$

In passing, we may note that the blood volume of the dog is about 9–10 per cent. of its body weight (or about 12 per cent. of the weight of the soft parts of the body), whilst the amount of water retained in the blood was reported by Engels as being only about 4 per cent. of that retained in the body. The inference is that the water which enters the blood is not distributed uniformly all over the system. As to where it goes Engels' analyses gave the following results (Table VI):

Table VI

Organ	Distribution of water
Muscles ...	482 gm.
Skin	126
Liver	21
Alimentary canal	16
Lung	14
Blood	11*
Kidney... ...	10
Brain	8
Uterus	2
Skeleton ...	0
	690

* In the blood at the end of the experiment; in addition to this, 20 c.c. of water had been removed during the bleeding, making 31 c.c.

Of the 776 gm., therefore, 690 were accounted for, leaving 86 unaccounted for.

The striking points are that the great bulk of the water went into the muscles and skin, which therefore are presented by Engels as the water depôts *par excellence* of the body. This is shown by Table VII:

Table VII

Weight of organ as a percentage of body weight	Organ	Weight of water stored as a percentage of total water stored
43	Muscle	68
16	Skin	18
41	Rest	14
100		100

Before accepting the suggestion embodied in the table given above, namely, that the skin and the muscles store water to an extent which is specific, the following fact must be considered. The skeleton and fat can scarcely be regarded as available store houses and together they amount to perhaps one-quarter of the body weight. Out of 75 gm. of available body, 43 would be muscle, 16 skin and 16 other organs. Then taking the weight of these as a percentage of the weight of body available for storage, Table VII would appear as follows:

Weight of organ as a percentage of available body	Organ	Weight of water stored as a percentage of total water stored
58	Muscle	68
21	Skin	18
21	Rest	14
100		100

and on that basis the distribution of water over the tissues of the body is much more uniform. While therefore it remains doubtful whether muscle and skin store water in virtue of any specific mechanism for the purpose, there remains in Engels' picture the very obvious feature that while the blood forms perhaps 10 per cent. of the weight of the dog, only about 2 per cent. of the water given is represented as remaining therein. In contradistinction to the picture presented by Engels, there is another and quite different one presented by Margaria (1930). Margaria (on a subject with a gastro-enterostomy, in which therefore the water went directly into the intestine) gave $1\frac{1}{2}$ litres of water. After allowing two hours for the body to attain osmotic equilibrium, Margaria obtained an increase in the water of the blood of about 48 c.c. per litre. The subject being rather small in stature, the total quantity of water retained in the blood was about 200 c.c. The whole quantity of water drunk was, as has been said, $1\frac{1}{2}$ litres, but of this more than half a litre had been eliminated and between 900 and 1000 c.c. retained. Thus the water in the blood amounted to about 20 per cent. of the water retained, though the blood itself probably amounted to only 7 per cent. of the body weight. This observation would make the blood an important store of water. Margaria's method of estimating the water was that of measuring the vapour pressure of the plasma by Hill's thermopile method (1930). It assumes, on the basis of Hill's own work, that the water in the blood was unbound.

It might be objected that the fundamental difference between the work of Margaria and that of Engels consisted in the fact that in the former case water was drunk and that in the latter salt solution was injected. There is yet another picture, given by Haldane and Priestley (1916), in which distilled water was drunk as in Margaria's experiment, but in which the result was apparently similar to that obtained by Engels; in fact Haldane and Priestley go farther than Engels in the direction of emphasising the absence of dilution of the blood. Between 10.45 a.m. and 1.15 p.m. Priestley drank five litres of water and between 10.45 a.m. and 3.15 p.m. he excreted 3205 c.c. of urine, so that on the balance he had retained approximately 1800 c.c. of water. As the last water had been drunk at 1.15, it may be supposed that the water had been practically all absorbed. If Priestley had a blood volume of 4·5 litres it will be seen that the water he retained was 40 per cent. of his blood volume. Yet the haemoglobin value of his blood was the same at the end of the experiment as at the commencement: indeed there never was any certain fall. We then have three pictures, which may be tabulated as follows (Table VIII):

Table VIII

	Observer	Material	Method	Apparent result
1	Engels	0·6 % NaCl injected into jugular vein	Haemoglobin estimations	Almost no dilution
2	Margaria	Distilled water drunk	Vapour pressure estimations	4·8 % dilution
3	Haldane and Priestley	Distilled water drunk	Haemoglobin estimations	No appreciable dilution

The first thing, perhaps, that strikes the reader is that the observations (1 and 3) in which the method of determination was the same give an identical result. The second is that all these determinations purport to deal merely with the percentage of water in the blood and not with the total quantity. We know nothing, in any of them, about either the constancy of the blood volume or the total quantity of haemoglobin. If, for instance, when Engels injected salt solution the spleen and other depôts had responded by adding to the number of corpuscles, more water might have remained in the blood than Engels supposed. And it must not be forgotten that 86 c.c. of water in Engels' experiments were unaccounted for. Similar considerations might be applied to Haldane and Priestley's experiments.

On the other hand, if in Margaria's experiment salts corresponding to 200 c.c. of plasma had migrated into the tissues from the blood, there might have been no extra water in the blood although the contents were diluted.

These two possibilities, the mobilisation of corpuscles and the migration of salts, are in no way mutually exclusive, and indeed they probably both take place to some extent, but we do not know to what extent. It seems desirable that an experiment should be carried out in which, after the administration of water, both the haemoglobin value and the vapour pressure of the blood should be determined on the same subject and at the same time. It is possible that in this event the results of both Margaria and Engels, contradictory as they seem, might be obtained.

When I commenced the writing of the present discussion I had supposed that the issue would lie as between free and bound water, that in some crude way we might regard the "binding" of water, *i.e.* the formation of hydrates, as "architecturally" the equivalent of the deposition of glycogen or of calcium, but as the discussion has developed the issue has receded farther and farther into the background. That is not to deny the possibility of "bound" water: it is merely to say that such evidence as the advocates of "bound" water bring forward has nothing to do with the storage of water which is superfluous to the basal quantity contained in the body. Summing up then so far, there appears to be little in common between the storage of water and that of carbohydrate, fat, or calcium. It is far from certain that there is any special or specific store, though in virtue of their bulk the muscles hold the major part and the skin a considerable quantity. Possibly, however, if epidermis and dermis were considered separately, and having regard to the proportionate weight of the epidermis, a case might be made out for regarding the dermis as a water depôt. Nor does there seem to be any chemical or physical line to be drawn between the water which is in store and that which is being in process of mobilisation. So far, the location of water storage has been considered with regard to the various organs of the body, but there is another angle from which it is worth a glance. Is water stored in the actual cells of the tissues concerned or in the spaces between the cells? If the reader is in sympathy with what has already been said, namely that the transfer of water is incidental to the osmotic and filtration conditions, he will naturally be prepared to discuss the possibilities of water storage, offered by the cells and the tissue spaces respectively, along those lines.

In this connection the work of Gamble and his colleagues in the Schools of Pediatrics in Johns Hopkins University and afterwards in Boston, is of great interest. It deals with the withdrawal of water from the stores.

An outline of their argument is as follows:

The water content of the juices secreted into the alimentary canal in the day is very great. Rowntree (1922) has pointed out that it may be several times that of the whole volume of water in the plasma. Most of this water is normally reabsorbed in the lower parts of the alimentary canal and therefore not lost to the body. When the gut is obstructed not lower than the duodenum this reabsorption does not take place and the body becomes dehydrated. Dehydration falls first upon the plasma, the plasma recoups itself in part at the expense of the tissue spaces. The tissue spaces do not, according to Gamble, recoup themselves at the expense of the cells. Gamble argues that the passage of water is merely incidental to the passage of ions. The stomach, for instance, secretes chlorine ions, hydrogen ions and some sodium ions, and the pancreas sodium and HCO_3 ions, and so forth. These materials are of about the same ionic concentration as in the plasma and the water goes out with them, the plasma in turn draws on the tissue spaces (presumably on account of the increased concentration of protein in the plasma) and sodium and chlorine ions pass from these spaces to the blood. But the tissue spaces cannot draw upon the cells in the same way, because the base in the cells is potassium not sodium and the potassium ions will not go through the cell membrane; if the chlorine ion goes through it gets replaced by the HCO_3 ion. For this reason the cell volume remains approximately constant and the tissue spaces form the body's principal store of water. Let us pause to look for a moment at Gamble and McIver's (1925) figures. In applying arithmetic to them I am conscious that I may seem to be endeavouring to get out of them more than the authors would expect them to bear. Clearly a very limited number of experiments is not susceptible of statistical treatment, and if I apply such treatment to them it is merely to clarify the issue and not to press home such deductions as may seem to result. I should like at this point to thank the authors for their kindness in personally placing the work at my disposal. Data from analysis of muscle and skin of rabbits No. 4 and No. 8, demonstrating differences in extent of possible availability of water from tissues, are given in Table IX:

Table IX

Values per 100 gm.	Muscle		Skin	
	Control (No. 4)	Pylorus obstructed (No. 8)	Control (No. 4)	Pylorus obstructed (No. 8)
Water	76·4	72·7	64·7	58·8

Therefore the muscle lost 3·7 gm. of water per 100 gm. of tissue and the skin 5·9 gm. per 100. Taking the muscle as half the body weight (roughly) and the skin as one-sixth, and the rabbit No. 8 to be 2 kg., the actual amounts of water lost are:

Muscle 37 gm.
Skin   19
        ——
        56

There were 225 c.c. of water found in the stomach of the rabbit, and it may be calculated from the controls, that 57 gm. would have been found in the stomach of a normal 2 kilo rabbit, under similar circumstances. The occlusion has therefore occasioned the presence in the stomach of $225 - 57 = 168$ c.c. of water. Of this only 56 c.c. have been accounted for by the loss from the muscle and skin respectively. But perhaps, as I have said, I am pushing criticisms too far.

Let us accept, provisionally, Gamble's general outlook, and from his standpoint look back to the experiments of Engels and Margaria respectively. It is clear that from Gamble's point of view the problems which Engels and Margaria present are fundamentally different.

Margaria gave *water* the distribution of which will be influenced by every electrolyte in the body; Engels injected *salt solution*, and had it been normal salt solution, the water might be expected to have gone no farther than the sodium chloride, that is to say, no farther than the tissue spaces. The ultimate fate of the water would depend principally upon the differences of hydrostatic pressure set up. It would in short find its way to those tissue spaces distension of which produces the smallest rise of hydrostatic pressure above that of the atmosphere. Of such tissue spaces those of the subcutaneous tissue are typical.

While therefore we cannot at present subscribe to the thesis that water is stored in the form of hydrates, its concentration both inside and outside the cell depends upon the attraction of other materials, chiefly electrolytes, for it, and its movements depend largely on the movements of the materials which thus bind it.

REFERENCES

BLOOR, W. R. (1922). *Physiol. Rev.* **2**, 107.

ENGELS, W. (1904). *Arch. Exp. Path.* **51**, 346.

FELL, H. B. and ROBISON, R. (1930). *Biochem. J.* **24**, 1905.

GAMBLE, J. L. and McIVER, M. A. (1925). *J. Clin. Invest.* **1**, 531.

HALDANE, J. S. and PRIESTLEY, J. G. (1916). *J. Physiol.* **50**, 296.

HILL, A. V. (1930). *Proc. Roy. Soc.* A, **127**, 9.

HUNTER, D. (1931). *Q.J. Med.* **24**, 393.

LEATHES, J. B. and RAPER, H. S. (1925). *The Fats; Monographs on Biochemistry*, pp. 143, 204. London.

LUSK, G. (1928). *The Science of Nutrition*, p. 321. 4th ed. Philadelphia and London.

MACLEOD, J. J. R. (1930). *Physiology and Biochemistry in Modern Medicine*, pp. 935, 936. 6th ed. London.

MARGARIA, R. (1930). *J. Physiol.* **70**, 417.

MAYER, A. and SCHAEFFER, G. (1913). *J. Physiol. Path. gén.* **15**, 510, 535, 773, 984.

—— —— (1914). *Ibid.* **16**, 1, 16, 23, 204.

MAYER, A. and TERROINE, E. F. (1907). *C.R. Soc. Biol.* **62**, 398.

ROBISON, R. and others (1923). *Biochem. J.* **17**, 286 and succeeding vols.

ROWNTREE, L. G. (1922). *Physiol. Rev.* **2**, 128.

STEWART, C. P. and PERCIVAL, G. H. (1928). *Ibid.* **8**, 283, 302.

TERROINE, E. F. (1914). *J. Physiol. Path. gén.* **16**, 212, 384.

—— (1919). *Physiologie des substances graisses*. Paris.

CHAPTER V

STORES II (OXYGEN, IRON, COPPER)

OXYGEN

Of the essentials for the preservation of life and function perhaps the most immediate is oxygen. It is remarkable that so necessary a substance should be stored to so slight an extent. Moreover, the higher up the scale of life, the more dependent the organism becomes upon an immediate supply of oxygen. The anaerobe can do without oxygen for ever, the frog for days, man for minutes.

Concerning the storage of oxygen, opinion has undergone a great change within my memory. Till almost the close of last century two principal facts stood out with regard to the physiology of living processes—the first of these was that every living form of activity which had been studied was normally associated with the demand on the part of the tissue for oxygen, the second was that the frog could live for a very long time in an atmosphere of nitrogen. It was known also that a frog's muscle would contract in a vacuum. The conjunction of these facts led to the doctrine of intra-molecular oxygen, then universally accepted, now only of historic interest.

The argument was clear, if (1) oxygen was necessary for contraction it must be present when contraction took place, (2) as contraction could take place with formation of CO_2 when the frog was in nitrogen, oxygen must be stored in the tissues of the frog, (3) as oxygen could not be extracted from these tissues by a vacuum, it must be held down by some chemical linkage and in that way it was given the name intra-molecular oxygen.

Perhaps the last authoritative textbook to teach the theory of intra-molecular oxygen was that of Schafer (1898 *a*).

Hermann's theory of muscular contraction assumes that the change is similar in kind to that which occurs on death, though less in degree. On death, he assumes that the hypothetical molecule which he calls inogen is split up into carbonic anhydride, sarcolactic acid, and myosin.... Hermann's theory just referred to was largely the outcome of his failure to discover oxygen among the gases of muscle—The oxygen used in the formation of carbonic anhydride must therefore be held in complex union within the muscle.

Oxygen then was stored in the organism, not as such but in some condensed or intra-molecular form.

The first serious blow to this theory was struck by Fletcher (1902): that seems to me to be a fair way of stating the case (if it is not I hope I may be forgiven). It is true, I believe, that Leonard Hill had previously done some work which pointed in the same direction, but neither the scope of the work nor the method of its publication gave it that stamp of authority necessary to overthrow a theory such as that of Hermann—a theory of universal acceptance based on accurate experiments and apparently reasonable argument.

The first point which Fletcher made by researches, which were as cautious as they were brilliantly conceived, was that muscular contraction did not necessarily produce carbonic acid. The inevitable conclusion was that if (a) neither the presence of oxygen as such, (b) nor the production of CO_2, was a necessary concomitant of the muscular contraction, there was no case for the view that the contraction process was essentially an oxidation. The background for the storage of oxygen in the intra-molecular form had completely vanished.

The whole of the subsequent work of the Cambridge school, Hopkins, Hill, Peters and their colleagues, has gone to support this view: but the strength of that work has not lain in its destructive but in its constructive aspect. The inogen theory has been replaced by another to which oxygen is no less necessary: indeed at the very time when Fletcher was dealing the death blow to intra-molecular oxygen, it was being shown, also at Cambridge, that oxygen was in the long run essential not only to contraction but to secretion and other living processes. The theory of muscular contraction which has replaced that of Hermann makes no less a demand upon oxygen; it is that the oxidation process is preliminary to, but not necessarily coincident with, the contraction of the muscle. If we compare the actual contraction to the release of a spring, the oxidation process is involved in the winding of the spring prior to its release.

In the frog the energy stored must represent the equivalent of quite a large quantity of oxygen. The classical experiment of Pflüger (1875) on the oxidation of living material showed that two frogs deprived of oxygen lived seventeen hours and gave off CO_2 at the normal rate for five hours and a half and at a somewhat diminished rate as time went on.

In the higher animals the amount stored is probably less in relation to the normal respiration of the organism but it would seem to differ in different tissues. The fact that the cerebrum leads a particularly hand-to-mouth existence, unconsciousness following as the result of a few seconds complete deprivation of oxygen, obscures the much more considerable power of other tissues to maintain some degree of functional activity under anaerobic conditions.

As regards stored oxygen as such, then, the only stores are:

(1) The oxygen in the gaseous state in the lungs (in the supplemental and residual airs), about 400 c.c. in man. In the fishes oxygen is stored as such in the swim bladder.

(2) The oxygen in physical solution in the tissue and body fluids which is not over 70 c.c. and may be much less—say, 10–15 sec. supply for the body.

(3) In the haemoglobin of muscles a doubtful but almost a negligible quantity, call it 50 c.c.

(4) In the haemoglobin of the blood, say 800 c.c., or about 2–3 min. supply for the body at rest.

Haemoglobin in man can hold, as we have said, 2 to 3 minutes' supply of oxygen, if the individual is at rest, and in so far as he moves he cuts down the length of time which oxygen so stored would last. It is probable that on such a basis haemoglobin has no survival value, and therefore its claim to be regarded as a store may be denied. In the case of certain lowly forms of life, quite different claims have been put forward, and on these claims we must dwell: behind them lies the general inference that originating as a store, the haemoglobin has taken on the more specialised function of an oxygen carrier. The controversy—if there can be said still to be a controversy—centres chiefly round the snail *Planorbis*, the larva of the fly *Chironomus* and certain worms such as *Arenicola*.

It was the contention of Lankester (1872) that "only its wealth of haemoglobin enables the *Annelid, Tubifex*, to live at the bottom of the polluted waters of the Thames".

The claim that the haemoglobin in the sac attached to the larva of *Chironomus* acts as a store was advanced by Miall and Hammond (1900), who showed that the larva could live an active existence for from 2 to 5 days in water which was freed from oxygen. The facts, particularly the freedom of the water from oxygen, were verified by Pause (1918), who concurred in the contention of Miall and Hammond that the larva lived on the oxygen stored in its own haemoglobin.

The general impression that the haemoglobin present in the circulatory systems of many invertebrate forms acted as an important store seems to have met with much favour until, in 1916, the matter was investigated in a quantitative way in Prof. Krogh's laboratory by Miss I. Leitch (1916). Miss Leitch worked on *Planorbis* and the *Chironomus* larva. In the case of *Planorbis* she showed in the first place that the whole haemoglobin of the snail only contains enough oxygen to meet its ordinary needs for about 3 minutes. So restricted a possibility of storage seems to bear no relation to the sort of claims put forward by Miall. Indeed it seems probable that in well-oxygenated water the haemoglobin of *Planorbis* blood exercises no function whatever. The quantity of oxygen which the circulating fluid can hold in solution at pressures above about 110 mm. suffices to satisfy the metabolic needs of the snail, and the life of that gastropod proceeds irrespective of the haemoglobin in its blood. The venous blood which comes from the foot is not reduced. The demonstration of the superfluity of haemoglobin under these circumstances lies in the fact that if carbon monoxide, sufficient to turn the haemoglobin into carboxyhaemoglobin be mixed with the air, the behaviour of the snail is unaffected. But if the partial pressure of oxygen in the air falls below 23 mm. a different story must be told. If then 1 per cent. of CO be mixed with the inspired gas the snail shows signs of deprivation of oxygen. In the case of the normal snails, "When the tension falls below

3 per cent. they remain steadily at the surface, moving little and breathing frequently". In the case of the snails which had been subjected to carbon monoxide

below 3 per cent. there is a difference. Here the snails become more and more sluggish, move very little and come to the surface slowly, at long intervals. At 2 per cent. they are not observed to come to the surface, and at 1 per cent. they are extruded from their shells and float about in the water....At tensions down to 7 per cent. *Planorbis* never makes use of its haemoglobin: the haemoglobin is always saturated. Further it does not require to use its lung, diffusion through the surfaces exposed to the water is sufficiently rapid to supply the needs of the animal. Below 7 per cent., however, the oxygen available by physical solution in the blood is no longer sufficient: the haemoglobin is reduced in the foot, that is it is constantly in use, and to accelerate the acquisition of oxygen the lung is frequently filled with air. It is clear that so long as the haemoglobin is saturated, there is no advantage to be had in renewing the air in the lung: but so soon as the blood becomes reduced it is of the greatest importance for its rapid oxidation that the lung with its rich net-work of veins and comparatively great facilities for diffusion should be repeatedly filled with fresh air.

Between 3 per cent. and 1 per cent. in the presence of haemoglobin as well as in its absence the snails become less active: but whereas by decreasing their activity, remaining at the surface and breathing with great frequency, they can with the aid of haemoglobin maintain a normal condition; in the absence of haemoglobin they are at 3 per cent. already incapable of doing so and go into a more or less latent condition. At a tension of 1 per cent. they cannot, without the aid of haemoglobin, supply even the small amount of air required in such a latent condition and float about in an obviously abnormal state. It is evident of how great importance to *Planorbis* the possession of haemoglobin is, since in the stagnant pools which it normally inhabits oxygen tensions from 7 per cent. downwards are common.

The condition of the snails when subjected to low oxygen pressures is very much like that in which we find ourselves normally. The quantity of oxygen which our circulating fluid can take up in solution is insufficient for our needs and the presence of a respiratory pigment is necessary to make up the deficiency.

The story with regard to *Chironomus* is much the same as that with regard to *Planorbis*. There is one difference in detail which is worth a moment's discussion. The dissociation curve of *Chironomus* "blood" is different from that of *Planorbis* (the latter has been confirmed approximately by Maçela and Seliškar (1925)) in that the oxygen is more tightly held in the case of the larval fly than of the snail. At a pressure of only one-fifth of a millimetre of oxygen the *Chironomus* haemoglobin is nearly half saturated. At this very low pressure therefore the haemoglobin could function to a quite important extent as an oxygen carrier. One-fifth of a millimetre oxygen pressure means roughly only 0·001 c.c. of oxygen per litre of water; with this fact in mind we may re-read Miall (1895)*.

So much for the claims of *Planorbis* and *Chironomus* to contain haemoglobin which acts as a store. Let us now turn to *Arenicola*. I must plead guilty of having suggested the possibility of the haemoglobin in *Arenicola* being used for the same purpose. In the case of the forms already considered the "store" theory is negatived by the simple fact that the oxygen capacity of the contained haemoglobin is small out of all proportion to the oxygen requirements of the animal over a period of time for which it is alleged that the organism can function in an atmosphere free from oxygen. The same is not true of *Arenicola*, and the suggestion with regard to it was as follows: between one tide and another *Arenicola* lives in a hole in the sand and the hour's supply of oxygen which is contained in its haemoglobin helps to make it independent of an external supply of the gas. The burrowing habits of *Arenicola* have been studied at Plymouth by Miss Mabel A. Borden, Ph.D., who has kindly given me permission to quote from her dissertation:

It is presumed that when the tide is in the worms have no difficulty in obtaining oxygen as they can either maintain a current through the burrow or come into direct contact with fresh sea water by moving to the surface.

During the period of intertidal exposure they are generally found at a considerable depth below the surface. It is probable that during this period which lasts for approximately three hours conditions approaching anaerobiosis will occur. The water in a complete burrow was observed in the laboratory suddenly to change the direction of its motion as the worm alternately protruded and withdrew its proboscis. Fine particles of sediment were observed in motion within the burrow. It is therefore assumed that the movement of the worm within its burrow will keep the water in a state of constant motion thus bringing it all into contact with the air at the surface. Since the openings are not more than 0·5 cm. in diameter the

---

* His statement is "Nevertheless the water was from the first exhausted of oxygen or nearly so". Was it exhausted of oxygen? or was it "nearly so"? That makes all the difference: but, as I said, Pause has looked into this question.

surface water exposed to air is obviously restricted and thus only a limited supply of oxygen can be dissolved in a given time.

The burrow is 1 to 2 feet deep. Miss Borden considers that this movement of the water in the burrow is the sole source of oxygen supply to which the *Arenicola* has access in the forms which she studied at Plymouth. The local conditions may be exceptional but they are none the less instructive. The *Arenicola* burrows in sand which is black except at the surface layer, which is brown. The blackness signifies the presence of reducing substances (sulphides), etc., which are black in the absence of oxygen, but brown on the surface when that gas is present. Thus it may be inferred that the sand in which the burrows are formed at Ballen Bay (Plymouth) is free from oxygen. Analyses of the water contained in the sand told the same story. Miss Borden devised an apparatus for extracting and analysing this water, but she never found in it more oxygen than was within the limits of experimental error of her method. That is to say, it was not more than in blank experiments in which thoroughly boiled water was manipulated in the same way as that obtained from the interstices of the sand.

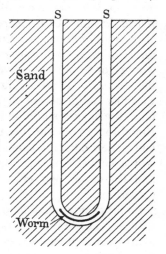

Fig. 52. Schematic view of the burrow of *Arenicola*. Superficial area of hole 0·2 sq. cm. Volume of water in each limb 12 c.c. Oxygen content of 12 c.c. of water 0·06 c.c.

The sand actually lining the burrow, it should be noted, was brown, deriving oxygen from the water in the burrow.

We are now face to face with the data required. An *Arenicola* is at the bottom of a U-shaped hole 60 cm. deep, at least 120 cm. long and ½ cm. in diameter. The worm requires about 0·22 c.c. per hour of oxygen or 0·6 c.c. for 3 hours. Can it obtain this by retracting and protruding its proboscis and so oxygenating the surface of the column of fluid 60 cm. away?

Each limb contains one-tenth of the requisite quantity on the assumption that the water is in gaseous equilibrium with air at the commencement of the three hours of "low water". On the assumptions which have been made, of which the most important are (1) that the interstitial water in the sand around the hole is free from oxygen, and (2) that the dimensions of the hole are about 120 cm. by 0·2 sq. cm. the worm must obtain a supply, if his needs are to be met, from the air 60 cm. away: 60 cm. is a long way for oxygen to diffuse through fluid in a tube. Roughton calculates that if the water were subjected to no disturbance by heat or movement, the time required for 0·6 c.c. of oxygen to travel from the surface to the *Arenicola* would be measured not in years but in decades. It travels at the rate of about 2–5 c.c. per century. Any small backward and forward movement caused by the movements of the proboscis would not be very effective in bringing the surface water down to the *Arenicola*. There are, however, some other considerations which we must weigh before we assume that the *Arenicola* is dependent upon the oxygen stored in its own haemoglobin. In this connection Roughton points out that the dimensions of the hole are such that convection currents cannot be left out of account. The influence which convection currents exert depends upon the temperature, and therefore their efficiency would appear to be variable and precarious.

So far as it is possible to sum up the evidence, the verdict seems to be as follows: if the *Arenicola* can make contact with all the water in its burrow it can secure but one-tenth of the oxygen necessary to satisfy its needs, and as the oxygen decreases the haemoglobin will no doubt exert a useful function in picking up the gas at low tensions and transporting it to the tissues; but the supply of oxygen appears at best much less than the occasion demands, and it has yet to be demonstrated that an hour's ration of the gas stored in the animal does not possess a survival value. The fact is, we do not know whether at the end of its low-water period the haemoglobin is or is not reduced in the animal generally. Doubtless we are right in viewing with great suspicion any theory which portrays the haemoglobin in the circulating fluid as having a function other than that of a vehicle of transport, and therefore I must admit that the responsibility of proof rests heavily on my shoulders if I am still to maintain that the oxygen in the blood of *Arenicola* can act as a store.

So much for such lowly forms as *Arenicola* and *Planorbis*, but there is quite another direction in which we may seek evidence of oxygen storage. The whale is said to remain under water for something of the order of half an hour; man can do the same for at most about 3 minutes. What mechanism enables the whale to perform this feat? The following data were given me by Mr A. Laurie, a member of the recent "Discovery" Expedition to South Georgia. The subject observed was a blue whale. The observations were made at Stromness.

Length	27 m.
Weight	122,004 kg.
Approx. weight of muscle	56,444 kg.
Blood volume (expressed in kg. but probably measured in litres)...	8,000 kg.
Lungs (an uncertain measurement)	1,226 kg.
Heart	631 kg.
Average oxygen capacity of 17 other whales was ...	213 c.c. per litre
Taking the oxygen capacity at 213 c.c. O_2 per litre of blood, that of the blood of the Stromness whale would be approximately	1,700 litres

How long might 1700 litres of oxygen be expected to last a whale the muscles of which weigh 56,444 kg.?

In making any estimate, the first thing noted is that the order of quantities is relatively much the same as for man: the blood volume is perhaps a little smaller in proportion to the body weight than it would be in man, but as it was measured by the "washing out method" it is not unlikely to be underestimated. The second point is a suggestion of uncertainty as to our right to suppose that the metabolism of the whale takes place on the same scale as that of man. Were one guided by the usual assumption that is made for the purpose of equating the metabolism to the size of animals, namely, that the metabolism varies with the superficial area, one would rate the metabolism per kg. of whale at a very low figure, so low in fact that the oxygen in its blood would last it about half an hour. It cannot be stated too clearly that we have no real knowledge as to whether the whale falls into the calculated place which would assign to it a metabolism per kg. of only one-tenth that of man. The whale lives in a very different environment and one in which a man would shortly die of exposure, that is, of the effects of too rapid conduction of heat from the body; on the other hand, the skin of the whale is very completely insulated. Moreover, there is this to be said on the experimental side. Mr Laurie tells me that he has made observations of the temperature of the whale after death. So limited is its power of losing heat that the corpse practically does not cool, *i.e.* the body temperature has not fallen appreciably before the commencement of chemical decomposition with consequent heat formation.

The intensity of the normal metabolism of the whale appears to be perhaps the most compelling problem in the, as yet, unascertained knowledge of the whale physiology. If that intensity were known, probably many other problems would be solved automatically.

Hitherto we have discussed the whale at rest, but other interesting problems arise when we consider the energetics of its movement.

The following is an extract from a letter to me written by Captain G. C. C. Damant, R.N.:

My friend H. H. Lodstone who is a professional hull designer with Thorneycrofts has worked out the attached. I gave him for data the dimensions of some sort of Rorqual from Beddard's book of whales and suggested that the form should be considered as torpedo shaped, perfectly smooth, and with the most perfect stream lining imaginable.

If the whale be SUBMERGED.

Length, 75 ft.

Diameter, 14 ft. maximum.

Sp. gr. 1·025 (*i.e.* cu. ft. per ton).

For the purpose of obtaining approximate volume, assume an ellipsoid of 70 ft. major axis.

Then vol. $= 70 \times 14^2 \times 0.5236$

                    $= 7183.79$ cu. ft.

Area of mean transverse section

$$= \frac{7183.79}{75}$$

$$= 95.7 \text{ sq. ft.}$$

Diameter of mean transverse section

$$= \frac{95\cdot7}{\frac{\pi}{4}} = 11\cdot03 \text{ ft.}$$

Surface $= 75 \times 11\cdot03 \times \pi = 2598\cdot8$ sq. ft.

The form of a whale is one in which eddy making is reduced to a minimum and may therefore be neglected.

If swimming BELOW THE SURFACE no surface waves will be produced and the only resistance to be overcome will be frictional.

Taking a coefficient of 0·01 we have for:

5 knots $\quad \dfrac{0\cdot01 \times 2598\cdot8 \times 5^{2\cdot83} \times 6080}{60 \times 33,000} = 7\cdot54$ E.H.P.

10 ,, $\quad \dfrac{0\cdot01 \times 2598\cdot8 \times 10^{2\cdot83} \times 6080}{60 \times 33,000} = 53\cdot46$,,

15 ,, $\quad \dfrac{0\cdot01 \times 2598\cdot8 \times 15^{2\cdot83} \times 6080}{60 \times 33,000} = 168\cdot08$,,

Effective horse-power is the theoretical thrust or tow-rope horse-power, to which must be added all the losses due to efficiency of propulsion.

If the whale is swimming AT THE SURFACE so as to produce surface waves, the horse-power will be increased to overcome this additional resistance as follows:

	5	10	15
Speed in knots	5	10	15
Length of whale in feet ...	75	75	75
$\sqrt{L}$	8·66	8·66	8·66
Displacement in tons	205	205	205
$\left(\dfrac{L}{100}\right)^3$	0·42	0·42	0·42
Displacement $\div \left(\dfrac{L}{100}\right)^3$	488	488	488
Resistance in lbs. per ton of displacement	1·5	30	170 approx.*
E.H.P. $= \dfrac{R \times D \times V \times 6080}{60 \times 33,000}$ ...	4·7	190	1,607
To which must be added the E.H.P. for frictional resistance as already calculated ...	7·5	53·4	168
Total Effective Horse-Power ...	12·2	243·4	1,775·0

* The displacement: length ratio being very high, at 15 knots it is "off the map" as far as my data go, hence the "approximate".

The points which the reader will notice are

(1) How small the power necessary is if the speed is slow.

(2) How considerable is the increase under any circumstances if the speed is sensibly increased.

(3) The enormous economy at the higher speeds of complete submersion when swimming. At 15 knots the whale if he is completely submerged uses less than one-tenth of the horse power that it would require if the surface were broken.

Let us now translate these figures into terms of oxygen. The equivalent of a horse-power is about 2·2 litres of oxygen. The figures given above then become:

Table X

Speed (knots)	Submerged			Surface		
	H.P.	O_2 per min. equivalent to H.P. (litres)	O_2 per min. assuming 20 % efficiency (litres)	H.P.	O_2 per min. equivalent to H.P. (litres)	O_2 per min. assuming 20 % efficiency (litres)
5	7·5	17	85	12·2	27	135
10	53·5	118	590	243	534	2,670
15	168	370	1850	1775	3905	19,525

Allowing 20 litres per min. for the basal metabolism of the whale, the graph shown in Fig. 53 gives an idea of the time which it might be able to stay under water at various speeds of travel. According to it it might amble along at 5 knots for more than a quarter of an hour without coming up to breathe, and in so doing would travel over a mile between respirations. Five knots as every small boat sailor knows will take you a long way if you can maintain it. It means leaving Dublin after breakfast and arriving in the Clyde before lunch the next day.

So much for the haemoglobin in the circulating fluid; but one of the most remarkable things about this remarkable substance is that, in some of the lowest forms of life in which haemoglobin is found, it appears in such places as nerve ganglia, muscle fibres, etc.

The physiologist cannot avoid the question—has this haemoglobin any function, and if so, what is it? Of such functions three at least may be considered:

(1) That it forms part of a system of transport.

(2) That it acts as a catalyst.

(3) That it forms a store.

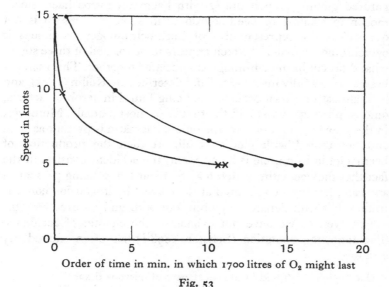

Fig. 53

There is, on the other hand, the possibility that haemoglobin in muscle and in nerve cells is simply an accident and has no significance whatever. Before proceeding to discuss the probable function of myohaematin it is necessary to enquire what reason there may be for supposing that it possesses a function at all.

If for the moment we must canvass the possibility of haemoglobin in muscle or nerve being an accident, it is at least allowable to enquire into the exact nature of the accident.

Here there seem to be two possibilities. One is that haemoglobin in small dilutions, too small for recognition, is in the circulating fluid and that, as carotene collects in the hen's egg, it is condensed in the red muscle or nerve. Of this there is no suggestion of proof. The other is that haematin in some form or other is of very wide distribution in tissues, and that the muscles in question have accidentally produced globin. The accident indeed would be a little more detailed. Globin appears to be capable of performing a rôle in either of two forms—the natured and the denatured. The denatured globin when it unites with haematin forms haemochromogen in alkaline reduced solutions: in some way which is not quite clear the representative of haemochromogen in tissues is cytochrome—a material which appears to be formed of three somewhat different haemochromogens associated together. This material has been very fully investigated and described by Keilin (1925), and is of almost universal occurrence, being found in aerobic bacteria, onions, porridge, yeast and the tissues of most animals. Moreover, in the growing point of, say, the onion, haematin is present as such and not united with globin at all. If then the production of haemoglobin in muscle is an accident, the accident consists in the fact that the haematin, which may be regarded as being present in any case, has been confronted at the correct hydrogen-ion concentration, not with denatured globin, but with undenatured globin.

But even so we have not exhausted the chapter of accidents. The following quotation (Schafer, 1898 *b*) has often excited my curiosity:

Haemoglobin diffused in the substances of nervous tissue:

(*a*) In the chain of nerve ganglia of *Aphrodite aculeata* (Lankester). In this annelid the chain of nerve ganglia possesses a bright crimson colour. The colour is most intense in the supraoesophageal ganglion, which has as intense a colour as a drop of fresh human blood. The colour impregnates the nerve itself and is not contained in the liquid bathing the tissues.

(*b*) An exactly similar observation has been made by Hubrecht who found haemoglobin in the red-coloured ganglia of certain Nemertine worms which possess no coloured corpuscles.

Mark you, "the colour is as intense as a drop of human blood". Is it also accidental that haemoglobin having once been formed should be concentrated in these nerve cells to the point of attaining a colour as intense as a drop of human blood? It may be!

As against this chapter of accidents must be put the general improbability of the wastage caused by the occurrence of accidental phenomena. It is quite a reasonable retort to the sceptic that the number of mechanisms in the organism which function is overwhelmingly greater than the number which do not, and that on any scheme of probability a mechanism must be assumed to have a function unless it is proved to be purely accidental.

Without riding this aspect of the case too hard, the probability of haemoglobin playing some sort of rôle where it is found in great concentration seems to warrant an enquiry into the possible rôles of which it may be capable. These, as we have said, appear to be those of an agent of transport—a catalyst—and a store. Let us therefore turn to them, and before deciding in favour of the last let us consider what is to be said in favour of the other two.

TRANSPORT

There is no reason to suppose that the haemoglobin of muscle or nerve actually circulates. It must, in the absence of information to the contrary, be supposed to be fixed in the cell. The rôle therefore which may be attributed to it in a transport system is that of a link in the oxygen-transport chain. Some such view as this has always been attractive. This view is supported by the recent work of R. Hill (1933), who has shown that the dissociation curve of myohaematin is almost if not quite hyperbolic in form and is so disposed under the circumstances under which he worked (pH 9·3 and 17° C.) that the muscle pigment attains a given percentage saturation at a much lower oxygen pressure than does that of blood. To quote his words: "It seems clear that the presence of muscle haemoglobin within the muscle cells will be of definite advantage in oxygen transport. . . . In the middle range of the dissociation curve there is a large difference in the relative saturations at equilibrium, which will allow the muscle pigment to take up the oxygen from the blood. . . . The muscle haemoglobin, with its relatively high affinity for oxygen, can be the intermediate carrier of molecular oxygen from the blood to the oxidase-cytochrome system in the cells".

CATALYSIS

But after all haemoglobin itself cannot be taking in oxygen, and giving it out at the same time. That the presence of haemoglobin can hasten the oxidation of certain fats has been shown by Miss Robinson (1924), but its power of doing so does not exceed that of other blood pigments, and in any case the scale on which it works scarcely seems to justify the assertion that it is present in muscle for that purpose. Whilst the evidence that haemoglobin in muscle performs any useful function as an oxygen catalyst is not satisfactory, one cannot use the word catalyst with regard to haemoglobin without some reference to the work of Henriques followed by that of Van Slyke and Hawkins and later of Brinkman and Margaria (1931). These authors seemed for a time to have shown that the pigment has remarkable powers of accelerating the velocity with which carbonic acid is converted into carbon dioxide and *vice versa*.

Henriques (1928) observed that when haemoglobin solutions, containing CO_2, were shaken *in vacuo*, a large part of the CO_2 was liberated into the gas phase very rapidly, whereas when plasma was similarly treated, the CO_2 evolution was slow at all stages of the processes. This effect was confirmed by Hawkins and Van Slyke (1930), who showed that it persisted even in diluted blood solutions. That the action is at any rate in large part catalytic is shown decisively by the recent work of Brinkman and Margaria, who found that the presence of dialysed haemoglobin even to the extent only of 1 part in 100,000 increased definitely the rate of the CO_2 reactions. These authors, like van Slyke and Hawkins, rather naturally attributed the catalytic effect to haemoglobin itself.

Yet here again haemoglobin faded out of the picture. More rigorous treatment by Meldrum and Roughton (1932) yielded a different explanation—namely, the existence of a specific enzyme, carbonic anhydrase, usually found alongside haemoglobin, which has the power of accelerating the reaction between CO_2 and water to form H_2CO_3 and also the splitting up of H_2CO_3 into CO_2 and water. Whilst the haemoglobin and the carbonic anhydrase are usually found together, haemoglobin may be found without the enzyme as in the blood of the worm, whilst the enzyme may be prepared free from haemoglobin. The blood of the foetus (in the goat) contains about half as much haemoglobin as that of the mother (volume for volume) whilst it only contains from one-tenth to one-hundredth of the quantity of enzyme.

The difference between haemoglobin as an agent of transport and as a store is merely the difference between space and time. As an agent of transport the haemoglobin takes up the oxygen at one place where the oxygen is, and gives it out at another place where the oxygen is not. As a store haemoglobin would take up the oxygen at one time when the oxygen is available and give it up at some subsequent time when the oxygen is not.

Now the particular interest about the possibility of haemoglobin as a store in the nervous system lies in the assumption that some neurones, and these important ones, can only function normally for a very short time if deprived of oxygen; deprive the brain of oxygen, and in less than a minute unconsciousness will supervene. It is reasonable therefore to calculate the benefit which might be derived from a store of haemoglobin in close relation to a nerve ganglion. Suppose the ganglion to weigh 0·1 gm., and suppose that one-tenth of its weight is haemoglobin ("which has as intense a colour as a drop of fresh human blood" was Schafer's phrase), the oxygen capacity of that haemoglobin would be 0·013 c.c.; suppose further that the remaining nine-tenths of the ganglion, or 0·09 gm., consumes oxygen at the rate of 0·03 c.c. per gm. per hour (the average rate of consumption of oxygen by the worm) or 0·0027 c.c. per hour for the ganglion; the haemoglobin on this computation could store 4 hours' supply of oxygen for the use of the ganglion. In making the above calculation no deduction has been introduced for the possibility of oxygen drifting from the haemoglobin in the ganglion to the surrounding tissues.

A matrix with haemoglobin appears so desirable for the nerve cells of animals which suffer great deprivation of oxygen, such as worms that burrow into soil charged with decaying vegetable matter, that one rather wonders why the arrangement is not universal. It is perhaps this surprise that deters me from laying more stress on the probability of haemoglobin in nerve cells performing the function of a store of oxygen.

In some sub-mammalian forms haemoglobin occurs in muscle, even in forms of life in which it is not found in the blood. When it does so, its appearance is usually rather spasmodic, picking out often a single muscle or set of muscles.

The muscles which contain haemoglobin are frequently of great importance, and I need only allude to the large literature in which the difference between red and white muscle is emphasised.

In vertebrates it is usual to find haemoglobin in the heart muscle. In the snails it is found in the muscles which operate the radula. In both cases therefore the haemoglobin is present in organs which perform constant spells of work. The mammalian heart could only live on the oxygen in its own haemoglobin for 10 seconds.

Until we know more about the physiology of death it will not be easy to decide whether 10 seconds' supply of oxygen in the heart itself has or has not any survival value.

But when we come to the consideration of red and white muscle in warm-blooded animals, I find it difficult to suppose that the haemoglobin in red muscle is important as a store, for the simple reason that I have given, namely, that the amount stored is so small relative to the amount used. An example which might possibly be cited as one in which the stored oxygen in red muscle with even a small supply of oxygen might have a survival value is the wing muscle of the bird. It is conceivable that some seconds' supply of oxygen in the wing might, in the case of some circulatory failure, prevent a crash.

One reservation may be made in the case of vertebrate muscle which functionally carries out rhythmic contractions, and which is typically red such as the heart or the wing muscles of birds. During the phase of the rhythm in which the muscle is in contraction the blood flow is much reduced or even stopped; during this phase the muscle fibre may receive its oxygen from the haemoglobin which surrounds it, the haemoglobin replenishing its store during the phase of muscular relaxation. Were such the case the haemoglobin would in a sense act as a store, but only for a very short time. The contemplation of such a process serves to illustrate what is in a sense the bane of such discussions, namely the ease of providing rather facile explanations of phenomena which we do not really understand. It is better to give no explanation than a wrong one. Nevertheless, if we exclude the rôle of oxygen storage from the functions of haemoglobin, we must recollect that many of the phenomena which have called forth the storage theory stand unexplained. With its fall let us return to *Chironomus*: the statement was that it could remain 5 days in water free from oxygen, could live an active existence under such circumstances and the larva could turn into the pupa. Altogether more convincing than the work of Miall is that of Cole (1921) on the conditions of life in Lake Mendota at certain seasons of the year. Lake Mendota appears to be an admirable place for such a research. The city of Madison which harbours the University of Wisconsin is placed on its shores. The situation of the Laboratories, but a few hundreds of yards from the water's edge, enables the terrain to be kept under constant observation, and the high level of scientific work for which that University is justly known among physiologists and biochemists gives additional confidence in Cole's researches.

The claim put forward by Cole and before him by Juday (1908) is that, at certain seasons, the water at the bottom of the lake is stagnant and is devoid of oxygen. Lake Mendota, which Birge studied with the greatest care, may be taken as an example. In the autumn, when the temperature of the lake has been nearly uniform at all depths, winds blowing across the lake set up currents which thoroughly "mix" the water and equalise the amounts of dissolved gases from top to bottom. When the ice covers the lake the mixing ceases and the water becomes more or less thermally stratified. Toward the spring, especially in the deeper parts of the lake, the water becomes stagnant, the oxygen being slowly used up by animals and decaying organic material. An excess of carbon dioxide accumulates in place of the oxygen (Birge and Juday, 1911). In spring, when the ice leaves the lake, the wind by circulating the water again equalises the amounts of dissolved gases and makes the lake habitable at all depths for aquatic animals.

As the season advances the water near the surface of the lake warms most rapidly and a stratification again takes place. Three regions are thus formed in late summer and early autumn, the hypolimnion, or stratum of stagnation and low temperature; the epilimnion or stratum of circulation and higher temperature (Birge, 1903), and between these, the narrow thermocline (Birge, 1903) or mesolimnion, characterised by a rapid transition in temperature. During the summer stagnation period, most of the animals in Lake Mendota migrate from the hypolimnion to the epilimnion. This was shown to be true for fish by Pearse and Achtenberg (1920) who set nets in the hypolimnion and rarely caught anything. The migration is due to the decrease of dissolved oxygen and the increase of carbonic acid. The same investigators showed that fish could not live for any great length of time in the stagnant region. They caught fish in the epilimnion, placed them in wire cages and then lowered the cages below the thermocline. They found that the fish died in about 2 hours, presumably of suffocation. However, there are animals in Lake Mendota which do not migrate from the stagnant deeper water. Some of these even live rather an active life throughout the stagnation period in the soft bottom mud and the water above it. Among such is *Chironomus*, on which Cole did the greater part of his experimental work.

In seeking to elucidate the source of the energy which *Chironomus* dissipates in surroundings apparently free from oxygen, Cole discards the suggestion of oxygen stored in the haemoglobin of the larva; he discards also the probability of a true anaerobic type of metabolism, and discusses two other alternatives, namely:

(1) That an enzyme complex, especially associated with the chitin of the envelope, is responsible for the formation of a peroxide which is broken down with the production of oxygen.

(2) That oxygen in the "atomic form" is constantly being made at the bottom of the lake in small quantities, that this atomic oxygen penetrates the epithelium which surrounds the vessels charged with haemoglobin, and that the gas so produced is then conveyed to the tissues.

The last view is the one to which Cole himself leans. Granting that the oxygen is produced as atomic oxygen, how does it escape becoming molecular oxygen in the water surrounding the larva? for remember that the haemoglobin of the larva is never in actual contact with the decaying vegetable matter. If, on the other hand, the oxygen is present in the water as ordinary oxygen, why was it not found by analysis? An answer might be that the quantity present was too small to be discoverable by analytical methods, but if that is once admitted the necessity for any special mechanism would disappear in view of the possibility that a quantity of oxygen, too small to be discovered by analysis, was present as the result of ordinary diffusion. What I think would be necessary in order to prove Cole's point would be the demonstration that while the water above the vegetable matter was free from oxygen, that in which it, and the animal life embedded in it, were soaked contained a measurable amount.

I have sometimes wondered why Cole so lightly set aside the possibility of the larva having a definitely anaerobic type of metabolism, *i.e.* one in which the respiratory quotient was infinity. Possibly the idea of a non-parasitic larva living an anaerobic life did not appeal to him, even though he quotes authors (Snyder, 1912, and Pütter, 1905) who think that an anoxybiotic existence was "the fundamental type and that the oxygen 'habit' was taken on during evolutionary development".

After all, it is not a very far cry from the larva of *Chironomus* to that of *Gastrophilus*. Kemnitz (1914) has shown quite conclusively that the metabolism of the *Gastrophilus* larva can oscillate without difficulty between the aerobic and the anaerobic types according as it finds itself in the presence or absence of oxygen.

There would seem then to be two views, firstly that the oxygen habit is fundamental and that anaerobic metabolism is acquired especially by parasitic forms of life, and secondly that anaerobiosis is fundamental while aerobiosis is a later development. About this matter I do not know enough to express an opinion, but at whatever stage aerobiosis has appeared, it seems that unless you can so far stretch language as to call the materials for anaerobiosis a form of oxygen storage, it is difficult to believe that oxygen is stored for any length of time in low forms of life.

As regards *Planorbis* and *Arenicola*, Miss Borden has studied a phase of metabolism which we have not yet considered, with a view to ascertaining whether the spells of oxygen-want which beset those forms can be tided over by its assistance. I allude to oxygen debt.

If *Planorbis* when using oxygen at a steady rate is deprived of that gas for an hour, Miss Borden finds that within the next 3 hours or so the snail will use up oxygen at a more rapid rate than the normal, so that over the whole period, including and subsequent to that of deprivation, the average oxygen usage for the period was up to the normal.

So far we have discussed haemoglobin as a store for oxygen, but while that issue is a matter of much uncertainty, there seems to be no doubt about the storage of haemoglobin for its own sake. A complete understanding of the storage of oxygen is not as yet. Within recent years, however, enough has been gleaned on the subject materially to alter the outlook.

IRON

It is currently stated that when the haemoglobin of the red blood corpuscles is broken down, the iron liberated is stored in the liver, ultimately to be rebuilt into haemoglobin. There is no reason to doubt the truth of this statement, but it is not very easy to give the sort of quantitative proof of it that is available with regard to the formation of glycogen in that organ. The proof rests principally on the appearance of increased quantities of iron in the hepatic cells after the injection of haemolytic agents. In certain cases of extensive liberation of haemoglobin, the storage of iron is not confined to the liver. The ferruginous material is found in the spleen and in other organs.

Comparing the storage of iron with that of glycogen, it is clear that iron is stored in a form simpler than that in which it is used; in the case of glycogen, however, the stored material is more complicated than the substance glucose from which it is made, but, from the standpoint of solubility, the two cases are similar and the transition from the circulating to the stored material is from the more soluble to the less soluble. So much may be said without prejudice to any ultimate judgment as to the precise form in which iron is stored; is it iron as such? or an oxide? or a metallic salt? or an organic compound?

It is usual to describe iron in tissues as "unmasked" or "masked" according to whether it can or cannot be stained with ferrocyanide without previous treatment of the section with acid. Again various words such as haemosiderin are used to denote the black deposits of material in which iron obviously plays a preponderant rôle. In these black deposits of haemosiderin the iron is "masked", though not infrequently a certain amount of unmasked iron is stained in the tissue surrounding the granules.

From the angle of storage it seems that haemosiderin is one form at least in which iron derived from the breakdown of haemoglobin is laid up. The complete proof of storage demands further that the iron is not being used and that it is held for future use. The views about haemosiderin, before the work of Cook, were very diverse, and varied from the conception of its being an iron-protein compound of the haemoglobin type, to that of its being "nothing other than elementary iron, overcast with a layer of oxide which gives the pigment its colour" (Fischer, 1924).

Sherburn F. Cook in 1929 published a very careful research on the composition of haemosiderin as found in the spleen of the horse. This organ is brown in section and very rich in iron, containing according to Cook as much as 4 per cent. of that metal. His findings may be quoted in his own words.

It is clear that haemosiderin is not a definite chemical entity, such as haemoglobin. The granules are of no particular shape or size but are analogous to red-blood corpuscles in that they form a substrate on which a chemical compound is deposited. Possibly they may carry the red-brown, iron-containing pigment in some such fashion as the corpuscles carry haemoglobin. At any rate the pigment may be removed, leaving the substrate, or stroma, intact.

This pigment is of peculiar nature. It contains no carbon or nitrogen and therefore cannot be an organic compound. Its iron content is very high, and aside from iron consists only of hydrogen and oxygen (or water). It must be an inorganic compound of iron, yet its reactions are not those characteristic of ferric iron. It gives none of the reactions of ferrous iron whatever. Hence this iron must be in the form of an oxide, or hydroxide. The behaviour of the material in solution suggests that it is in the colloidal state. Furthermore, the behaviour of a pure ferric oxide sol. towards acids, and towards ferricyanide and thiosulphate, has been shown to coincide to a reasonable degree with that of the pigment. If this is so, then we may tentatively define haemosiderin as some form of colloidal ferric oxide, physically combined with an organic substrate, the stroma of the granule. It may be adsorbed on the surface of the granule but it is more likely that the latter is permeated by the iron compound, the molecules being held in place throughout the substance of the granule by physical forces. Were the iron held on the surface it would be quickly removed by strong acid, but observation has shown that some of it persists in the interior and can there become blackened by sulfide after the surface is no longer affected by that substance. Another possibility is that the iron in the interior of the

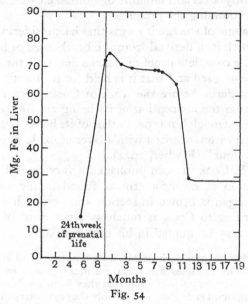

Fig. 54

granule is in the form exhibited in the pigment but that the surface is covered with a thin layer of free ferric iron. This conception would coincide to a certain extent with that of Fischer. But there seems to be no metallic iron present.

The nearest thing which we possess to prove that iron once stored is subsequently used is provided by the circumstances of early mammalian life, pre- and post-natal. Bunge (1902) many years ago was responsible for the general conception that the child at birth contained a sufficient store of available iron to supply its needs until after it was weaned, there being an insufficient amount of iron in the milk. Delicate tests for iron have served to modify this conception only in detail. The recent observations of Ramage, Sheldon and Sheldon (1933), however, fit in exactly with Bunge's view. By spectroscopic methods they have estimated the amount of iron in the human liver, from the 24th week of intra-uterine life onwards (Fig. 54). The total quantity rises rapidly before birth, falls slightly at first, and about 12 weeks after birth drops suddenly to less than half its value at birth. There it remains for some time, and does not regain its value at birth until the baby is about 60 weeks old.

COPPER

Copper appears to be stored in the liver. This is perhaps most evident in foetal and early post-natal life. When I say "most evident" I mean merely that the demonstration is then more easy. As has been said in other connections, the complete demonstration should consist in the facts (*a*) that copper is found in the liver, (*b*) that at some period of its residence in the liver it is not being used, and (*c*) that it is being held to be used for some demonstrable purpose later.

In foetal life as the course of pregnancy proceeds, the percentage of copper and, *a fortiori*, the total quantity of that element in the liver increases. In fact the copper in the liver and the iron go hand-in-hand. The amount of copper at any date may be within about 20 per cent. of what would be calculated if the figure for the iron given in Fig. 54 were divided by ten.

After birth the copper in the liver diminishes until the time at which the infant ceases to subsist upon milk; after that, the percentage of copper remains fairly constant but the total amount present increases with the size of the liver. So far then there is no difficulty in demonstrating the presence of copper in the liver.

The proof that it exists there functionally in a state of suspense appears a matter of inference, and in so far the complete demonstration of its storage fails.

On the other hand, the purpose to which it is put furnishes the motif for a very fascinating story. Copper seems to be bound up with the formation of haemoglobin. This was shown by Steenbock and his co-workers (1925) in a series of now classical researches. The general trend of these was as follows:

(*a*) Young rabbits, rats, etc., after the age of weaning, could be made anaemic by a diet of milk.

(*b*) Green food in addition to the milk abolished the anaemia.

(*c*) Iron carbonate did not abolish the anaemia.

(*d*) Further addition of an alcoholic extract of this green food was, however, efficacious.

(*e*) Not only was the extract of green food effective but also the ash of the extract.

(*f*) Ultimately the active principle in the ash proved to be copper.

Copper is not absent from milk, and there is independent reason for the belief that some forms of infantile anaemia are due to a deficiency of copper, *i.e.* that the copper stored in the liver during foetal life, plus the copper ingested in the milk, are together inadequate for the needs of the growing child. In this connection this subject has been studied by Gorter, who points out that milk from the breast is about twice as rich in copper as cow's or goat's milk.

Milk from	Breast	Cow	Goat
Copper in mg. per litre	0·24	0·12	0·15

Incidentally the practice of diluting cow's milk further increases the disparity.

With regard to nutritional anaemia in infants, Prof. Gorter has kindly allowed me to quote a lecture, the manuscript of which is in the press:

This type of anaemia is more often seen in twins and in prematurely born children. The age is mostly six months to two years. It is of rarer occurrence after breast feeding than when a child has got cow's milk. I agree with Scheltema that goat's milk produces anaemia with greater facility than cow's milk. It is probably more frequent in rural districts than the town. In type, alimentary anaemia is mostly an oligosideremia, the number of erythrocytes being proportionally less reduced than the amount of haemoglobin. Cure of alimentary anaemia is easily obtained by adding vegetables or fruit or meat to the diet, as well as iron preparations like lactate, or oxalate of iron.

The facts then that lack of copper produces anaemia and that certain forms of anaemia may be cured by the administration of copper-containing substances seem to indicate that copper plays a rôle in the formation of haemoglobin; what rôle we do not know. The evidence is against the inclusion of a copper atom as one of the normal constituents. Haemoglobin completely free from copper has, I believe, never certainly been prepared from blood; but specimens have been prepared in which the content of copper is as low as 0·019 mg. of copper per gm. of dry haemoglobin (Elvehjem, Steenbock and Hart, 1929), *i.e.* much less than one atom of copper to a molecule of haemoglobin (mol. wt. 68,000). Copper clearly does not enter into haematin as prepared by Hans Fischer (1929). Much more likely is it a catalyst of some kind which promotes the formation of haemoglobin without entering into the composition of that pigment; such a view would not be out of keeping with the work of Elvehjem (1931), who showed that the presence of copper in small quantities not only increased the rate of growth of yeast, but also the production of the *a* component of cytochrome. "Copper", in the words of Elvehjem, "has the property of stimulating the formation of certain haematin compounds."

REFERENCES

BIRGE, E. A. (1903). *Trans. Amer. Micr. Soc.* **36**, 223.
BIRGE, E. A. and JUDAY, C. (1911). *Wisc. Geol. Nat. Hist. Soc. Bull.* **22**, x + 529.
BORDEN, M. A. As yet unpublished.
BRINKMAN, R. and MARGARIA, R. (1931). *J. Physiol.* **72**, 6 P.
BUNGE, G. (1902). *Text-book of Physiology and Pathological Chemistry*, p. 337. 2nd Engl. ed. London.
COLE, A. E. (1921). *J. Exp. Zool.* **33**, 293.
COOK, S. F. (1929). *J. Biol. Chem.* **82**, 585.
ELVEHJEM, C. A. (1931). *Ibid.* **90**, 111.
ELVEHJEM, C. A., STEENBOCK, H. and HART, E. B. (1929). *Ibid.* **83**, 21.
FISCHER, H. (1924). *Oppenheimer's Text-book*, **1**, 368. 2nd ed. Jena.
—— (1929). *Annalen Chem.* **468**, 98.
FLETCHER, W. M. (1902). *J. Physiol.* **28**, 474.
GORTER, E. As yet unpublished.
HAWKINS, J. A. and VAN SLYKE, D. D. (1930). *J. Biol. Chem.* **87**, 265.
HENRIQUES, O. (1928). *Biochem. Z.* **200**, 1.
HILL, R. (1933). *Nature*, **132**, 897.
JUDAY, C. (1908). *Trans. Wisc. Acad. Sci. Arts, Lit.* **16**, 10.
KEILIN, D. (1925). *Proc. Roy. Soc.* B, **98**, 312.

KEMNITZ, V. (1914). *Verh. Deutsch. Zool. Gesel.* **24**, 294.

LANKESTER, R. (1872). *Proc. Roy. Soc.* **21**, 70.

LEITCH, I. (1916). *J. Physiol.* **50**, 370.

MAÇELA, I. and SELIŠKAR, A. (1925). *Ibid.* **60**, 428.

MELDRUM, N. U. and ROUGHTON, F. J. W. (1932). *Ibid.* **75**, 15P.

MIALL, L. C. (1895). *Natural History of Aquatic Insects*, p. 133.

MIALL, L. C. and HAMMOND, A. R. (1900). *The Structure and Life History of the Harlequin Fly.* Oxford.

PAUSE, J. (1918). *Zool. Jahr.* **36**, 339.

PEARSE, A. S. and ACHTENBERG, H. (1920). *Bull. U.S. Bur. Fish.* **26**, 293.

PFLÜGER, E. (1875). *Pflügers Arch.* **10**, 251.

PÜTTER, A. (1905). *Zeitschr. allg. Physiol.* **5**, 566.

RAMAGE, H., SHELDON, J. H. and SHELDON, W. (1933). *Proc. Roy. Soc.* B, **113**, 308.

ROBINSON, M. E. (1924). *Biochem. J.* **18**, 255.

SCHAFER, E. A. (1898 a). *Text-book of Physiology*, **1**, p. 110. Edinburgh.

—— (1898 b). *Ibid.* **1**, 187.

SNYDER, C. (1912). *Sci. Prog.* **6**, 107.

STEENBOCK, H. and his co-workers (1925). Papers in *J. Biol. Chem.* **65**, and succeeding volumes.

CHAPTER VI

STORES III (BLOOD)

The conception that certain organs can play the rôle of storehouses or depôts for blood has attained considerable prominence during the last few years. It is desirable to review the experimental evidence on which this claim is based. The organs in question are the spleen, the liver, the lung and the sub-papillary vessels of the skin. To these may perhaps be added the veins of the pregnant uterus.

SPLEEN

Above I said that the conception of blood depôts "has attained considerable prominence" lately, rather than that "it is new", for a century ago Hodgkin (1822) regarded the spleen as a reservoir of blood and 32 years later Gray (1854) took the same view; it is difficult in the light of modern work to say why that view dropped completely out of sight. Possibly it was prejudiced by Gray's own denial to the spleen of unstriped muscle. The presence of unstriped muscle in the trabeculae of many spleens is of course strong support to the view that the organ is capable of great alterations in size, but even in the absence of a muscular integument it is evident that the organ might be filled and emptied of blood by a suitable manipulation of the muscle in the splenic artery and vein.

My own introduction to the subject is discussed later in this chapter. The first indirect proof that the spleen could contain blood which was outside the general circulation was furnished in 1923 (Barcroft, J. and H.); it was that CO inhaled by the resting animal only very slowly entered the blood in the spleen pulp, slowly, that is, compared with the rate at which it entered the general circulation and the blood of other organs; this result was verified and worked out in detail by Hanak and Harkavy (1924). Here again, the fact that in cases of death from CO poisoning the haemoglobin of the spleen pulp was relatively free from CO was observed by Heger (1894), but neither by him, nor by others was the obvious inference drawn.

Another indirect method is that used by Scheunert and Krzy-wanek (1926), Binet and his school (1926), Izquierdo and Cannon

(1928) and by others; it is to compare the reaction of the animal before and after splenectomy. Personally, I have a predilection for direct methods, and therefore I have approached the subject in a no more elaborate way than that of observing the size of the spleen when the organism reacts. This may be done in the intact animal in the following way. In a preliminary operation metal clips are placed round the edge of the spleen. The animal is radiographed in two perpendicular planes of space (*i.e.* from the side and from the dorsal aspect): it is then possible from the radiograms to construct a model which represents the superficial area of the spleen (Barcroft, 1925; Barcroft, Harris, Orahovats and Weiss, 1925).

A simpler way of making such observations is to exteriorise (Barcroft and Stephens, 1928) the spleen, in which case, so marked are the alterations in volume of the organ, that the most casual observation cannot fail to discover considerable alterations in its size. From such alterations may be inferred the expulsion of the stored content of corpuscles. The spleen of a dog of about 18 kg. is capable of holding at least 300 c.c. of blood, *i.e.* roughly 2,000,000,000,000 red corpuscles, which can be expelled in case of necessity into the circulation. The number of corpuscles in the dog would be about 9,000,000,000,000, so that approximately one-fifth of the corpuscles of the dog may be held in reserve in the spleen. The occasions on which the spleen vents itself may be classified as those in which there is immediate need either for a blood richer in corpuscles, or for an increased blood volume. The line between the two is probably not a very hard and fast one. It is probable that blood becomes richer in corpuscles in the spleens of those animals which have a very muscular capsule, the rhythmic contractions of the muscle functioning to some extent as a filter press and squeezing out lymph. In the horse, the dog, the cat, etc., the blood coming from the spleen will be rich in corpuscles, as indeed we know it to be (Cruickshank, 1926), whereas in man the significance of the splenic contraction will be entirely an augmentation of the volume of blood in circulation. Polycythaemia occurs during exercise in splenectomised no less than in normal individuals.

Perhaps the most obvious demand for an addition to the volume of blood in circulation is at the onset of physical exercise (see

Chapter VII), and the disgorgement of the spleen during exercise has been shown by all of the methods described above. A demonstration, however, which always makes a strong appeal to me, is that of Capt. G. R. McRobert, I.M.S. (1928), of Rangoon. McRobert takes pairs of rats, each member of the pair being from the same litter and as alike as possible. He divides the rats into two groups *A* and *B*. Each pair is represented in group *A* as also in group *B*. The *A* rats while in full exercise, swimming, are killed instantaneously. The *B* rats are stalked in their sleep and hit on the head without awaking. Fig. 55 shows a typical example of the spleen of each.

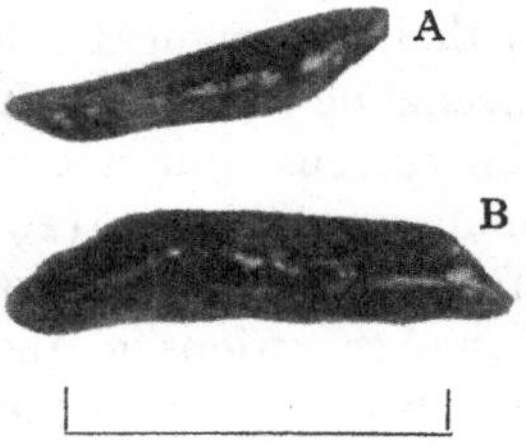

Fig. 55. Spleen of male rats, litter brothers. *A*, exhausted; *B*, resting. (McRobert.)

Acute conditions, other than exercise, in which there is a marked expulsion of blood from the spleen, are haemorrhage, oxygen want, mental agitation, and the influence of anaesthetics. Sub-acute ones are pregnancy, post-operative conditions, and possibly lactation.

(1) *Haemorrhage.* When an animal is bled, the spleen contracts, and it is possible in a crude way to measure the volume shrinkage. Hence a comparison may be instituted between the amount of blood taken out of store, or at least out of that particular store, and the amount of blood lost. Such a comparison reveals the fact that in the early stages of haemorrhage the blood volume scarcely diminishes at all, the spleen making up immediately almost what the animal loses as the result of the haemorrhage. The following figures were obtained from a dog:

Successive bleedings

	(1)	(2)	(3)	(4)	(5)	(6)
A	20	40	68	133	200	230 c.c.
B	5	25	39	44	50	52 c.c.

And also from a cat:

A	10	37	61 c.c.
B	9·8	16·5	20·2 c.c.

A = Total quantity of blood lost by animal.
B = Total quantity of blood lost by spleen.

(2) *Emotional conditions.* The effects of emotion on the spleen were first demonstrated by Hargis and Mann (1925), who constructed a type of plethysmograph in which the spleen could lie without apparent discomfort to the fully conscious animal. Events such as the closing of the door, or any sudden noise, caused contraction of the organ. Izquierdo and Cannon (1928) have pointed out that such conditions as fear and rage, those in fact in which adrenaline is secreted, produce a marked increase in the red blood corpuscle count in cats which contain spleens, but on the other hand, in splenectomised cats no such increase takes place. Previous to their experiments, Binet and his pupils (1927) observed that adrenaline itself raises the red blood count in animals which contain spleens, but not in those which have been splenectomised. Feldberg and Lewin (1928) have correlated the observed contraction of the spleen with the increase in the red blood count. So far as my own observations go, the emotional conditions which I have studied have been such as precede exercise, or are calculated to precede it. In this connection an attempt (Barcroft, 1930) was made to grade the emotional stimuli, and to discover what grade of stimulus evoked the response (i) of the spleen, and (ii) of voluntary muscle. These experiments were carried out on a dog in which the desire to pursue any cat in the vicinity was rather, but not too strongly, marked.

(*a*) If this dog were laid on the table and a clean duster dangled in front of its nose, no notice was taken of it, and the size of the spleen remained unaltered. If for that duster another was substituted which had been in a basket where there was a cat, there was an appreciable though not large contraction of the spleen. There was no gross muscular response in the sense of effort to move. It is not possible to say whether there was heightened tone of the skeletal muscles (Fig. 56*b*).

(*b*) If the cat, being in an adjacent room, were caused to "mew", a further but still incomplete contraction of the spleen would take place (Fig. 56*b*). The muscular movement associated was at most pricking up of the ears, and movements of the eyes and head.

(*c*) The cat was then brought into the room and placed where the dog could see it, and about 2 feet away. This procedure was associated with a further contraction of the spleen and muscular restlessness on the part of the dog (Fig. 56 b).

(*d*) Finally the dog was told to chase the cat, which it was allowed to do for about a quarter of a minute, with the result that its spleen contracted still more. In one case, that of June 5th, the contraction was probably almost maximal, at all events the surface of the spleen (13·5 sq. cm.) approached closely to that in the animal post-mortem (13 sq. cm.). This dog was killed on October 30th following. Table XI gives a numerical statement of the effects observed (see also Fig. 56 a).

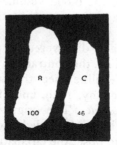

Fig. 56 a

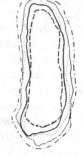

Fig. 56 b

Fig. 56 a. R, rest; C, dog sees cat. The numbers represent the relative sizes of the dog's spleen.

Fig. 56 b. — · — rest; —— smells cat; ······ hears cat; —— sees cat; ------ chases cat.

Table XI. *Approximate volume of dog's spleen expressed as a percentage of the volume at rest before the experiment*

Date	Rest	Smells cat	Hears cat	Sees cat	Chases cat
Mar. 29th	100	—	64	—	26
Apr. 2nd	100	—	—	45	28
Apr. 3rd	100	—	—	46	30
Apr. 7th	100	—	—	34	25
May 4th	100	72	51	39	—
June 5th	100	62	54	33	13

The figures suggest a better response on the part of the spleen at some times than at others. The measurements on which they are based are, however, too rough for any stress to be laid on this point.

If no more than the emotions were involved there would be little point in the expulsion of a considerable quantity of blood from the store into the circulation. The significance appears to be that the series of emotions enumerated ultimately lead up to an effort of physical exercise, and that the spleen has anticipated the demand for that increase in the blood volume which is part and parcel of the complex which insures augmentation of the minute volume of circulating blood.

The principal facts discovered on animals have been reported on human beings, i.e. the contraction of the spleen with exercise, emotion and adrenaline (see the work of Benhamou, Jude, Marchioni and Nouchy, 1929, and of Castex and di Cio, 1929).

(3) *Carbon monoxide.* The original discovery that the spleen acted as a store of blood depended upon the fact that CO, given in doses insufficient to upset the circulation, only reaches the haemo-globin of the spleen pulp very slowly (Barcroft, J. and H., 1923). But if the gas be administered in greater concentrations it produces a contraction of the spleen with consequent expulsion of the splenic contents (Hanak and Harkavy, 1924). The mechanism of this con-traction was studied by de Boer and Carroll (1924). They showed that if the spleen was connected with the body solely by the splenic nerve, the spleen itself being perfused from an artificial circulation, the administration of CO to the animal (none of whose blood entered the spleen) produced contractions of the organ, whilst the con-version to CO-haemoglobin of a considerable portion of the haemoglobin in the blood which traversed the spleen did not affect the volume of the organ. The effect of the CO is on the nervous system, and it can act effectively on the spinal cord.

The expulsion of blood from the spleen on administration of CO furnishes an example of the fact that the possession of a store of blood may, under certain chosen circumstances, determine the difference between life and death. Three groups of guinea-pigs were exposed to carbon monoxide in lethal doses. Group I, normal; Group II, splenectomised; Group III had been subjected to an abdominal operation which did not involve the spleen. Of the three groups, the second died in the carbon monoxide chamber before the members of Groups I and III; had these been removed from the chamber just after Group II died, they (i.e. I and III) would have survived, and would have owed their survival to their spleens.

(4) *Certain anaesthetics.* No phenomenon in connection with the volume of the spleen is more striking than the shrinking of this organ when chloroform and ether are administered either separately or together. The smell of these anaesthetics will cause contraction of the spleen in many dogs; that, however, is clearly a different phenomenon and, in so far as it is of interest, it should have been mentioned among the reactions of the spleen to emotion. On the administration of anaesthetic doses, the spleen usually contracts to a very small bulk, expelling its contents into the circulation. It may be of some comfort to the surgeon to reflect that when he is operating (certainly this is so in the dog, cat, etc.) there is an abnormally large quantity of blood in the vessels. Indeed the function of the spleen as a blood depôt would probably have been appreciated much sooner but for the fact that on the operating table the surgeon usually sees it more or less empty. Surgeons have told me that under spinal anaesthesia they have seen it in its distended condition.

Alcohol produces no such reaction: the only shrinkage of the exteriorised spleen which I have observed in dogs under the influence of alcohol seemed to me to be associated entirely with general excitement. During the condition of complete unconsciousness there is not in the case of alcohol, as there is in the case of ether, a contracted spleen. Interesting as are the conditions which evoke passing contraction of the spleen, the causes of long-continued contraction are not less interesting. The best established of these are pregnancy and heat (Barcroft and Stephens, 1928), and certain post-operative conditions.

(5) *Pregnancy and heat.* Concerning pregnancy, it is evident that in the later stages there is a quite unusual volume of blood in the generative organs. It is not very easy to measure the blood in the generative organs or to correlate it with any other variable. Barcroft and Rothschild (1932) made determinations on rabbits, cats, and dogs, in which animals the most obvious correlation was that between the quantity of blood in the uterus and the total weight attained by the foetuses (Fig. 57). Here it need only be said that in all these types, in spite of the difference in the size of the adult animal, the amount of blood in the generative organs is commonly in the

neighbourhood of 30–35 c.c. at the end of pregnancy, and after parturition the quantity rapidly diminishes to something negligible. This quantity of blood, in the case of the dog at all events, seems to be transferred from the spleen—a very intelligible arrangement, when it is recalled that, from the metabolic point of view, the biochemical situation in the mother is no doubt being exploited for the provision of blood in the rapidly growing embryos. Thus the foetuses within the uterus of a bitch of 18 kg. may weigh over 2 kg., and may con-

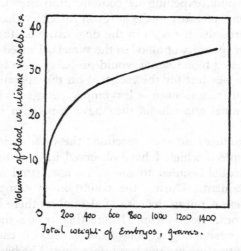

Fig. 57. Relation of quantity of blood in uterine vessels of bitch to total weight of contained embryos. Probably correct to within 25 per cent.

tain perhaps 200 c.c. of blood. Most of this must be made within a month from materials supplied by the mother. It is not surprising, therefore, that the mechanical needs of the hypertrophied maternal organs should be met from the store in the spleen. The contraction of the spleen in the dog is usually well marked during heat, when the generative organs become turgid; when heat passes off, the spleen resumes its normal size, and contraction sets in again about half-way through pregnancy, coming on gradually and reaching its most accentuated condition about 2 days before the pups are born (Fig. 58).

A word here concerning the mechanism of this pregnancy contraction. Is the transference of blood from the spleen to the uterus the result of a hormone, or must it be attributed to the action of the nervous system? So far as our meagre knowledge goes, the

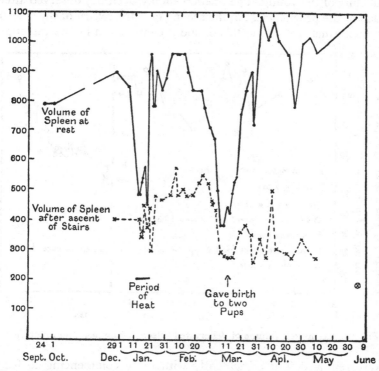

Fig. 58. Computed volume of spleen during heat and pregnancy: at rest and after standard exercise; ⊗ after chasing cat along passage. The mass of the spleen in grams is not very different from the figure obtained by dividing the arbitrary units on the ordinate by fifteen.

latter is correct. A bitch, the spleen of which had been carefully denervated, has been observed now through two pregnancies. Neither at heat nor during the pregnancy was there any marked contraction of the spleen. So far as we know, therefore, the pregnancy contraction of the spleen is a manifestation of the activity of the nervous system. The reaction noticed in this animal was a rather curious one, namely, that the spleen contracted, not during, but after pregnancy; whether it was a coincidence or not, the post-partum contraction corresponded with the period of suckling of the young. The phenomenon was clearly observed after each of the pregnancies; in the second case the return of the spleen to its normal size was watched, and it commenced before the pups

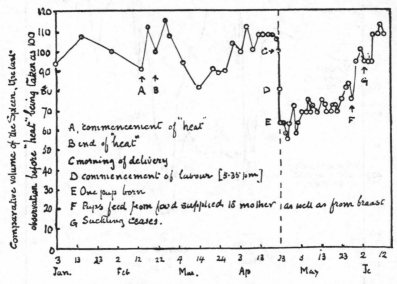

Fig. 59. Computed volume of denervated spleen during heat and pregnancy.

were removed, but coincidently with these commencing to feed from their mother's dish (Fig. 59)*.

(6) Of the instances of splenic contraction which persist over weeks, none is more striking than that following operative procedure. The number of recorded cases is not large, but they are very remarkable. During the course of experiments in which parts of the intestine were exteriorised, some time after exteriorisation of the spleen,

* No reliable post-mortem examination was made on this animal, but the spleen of another, denervated the same way, was found to have at least one trabeculum which was not completely denervated.

observation revealed some quite unexpected facts. In the first two experiments of the series a loop of small intestine was withdrawn and placed outside the abdominal wall. In both cases the animal seemed to do well, even though the gut had not been severed, and the food had to pass along the exteriorised portion. Neither dog, however, lived over 4 weeks, the ultimate trouble in each case being perforation. In one animal the lesion was in the abdominal cavity; the dog became suddenly ill and the condition clearly being hopeless, it was killed. The post-mortem examination showed a general infection of the abdomen.

The perforation in the other animal took place in the exteriorised portion; it was due to an accidental cause, namely, that the dog swallowed a piece of bone about 4 inches long. This was unable to get round an angle where the gut was fixed to the abdominal wall. In both these cases the spleen became quite small during the operation, and did not regain the size thereafter. Indeed, in the first case, that in which general peritonitis developed, the spleen had shrunk even more than in the latter, in which only perforation took place; it was at its smallest when the animal died. The question arose: Had these animals survived, would their spleens ever have attained to a size as great as that which they would have occupied if the subsequent abdominal operation had never been performed?

One experiment has taken place in which a portion of gut has been exteriorised with perfect success (Barcroft and Florey 1929) 2 months after exteriorisation of the spleen. There was no indication of any general or even local infection. Nevertheless, the spleen shrank immediately on the intestinal operation to about half the previous size, and remained so for about 4 weeks. After that time it dilated, and attained the size which, judging from a control dog, it would have occupied if no intestinal operation had taken place. Here again, we are in the dark as to whether the contraction was due to a nervous reflex or to some substance absorbed from the wound or elsewhere. A less severe operation is that of opening the abdomen, exposing the intestines under aseptic conditions, putting them back and stitching up the abdominal wall. Even this produces a contraction of the spleen which lasts for some days. The following (Table XII) are approximately the relative

Table XII

Day after exteriorisation of spleen	Approximate relative volume of spleen	Remarks	Temp. °F.
116	108	—	—
123	93	—	—
140	{ 100	Before operation	—
	{ 66	After operation	—
142	69	—	100
144	66	—	100
148	83	—	100
149	—	—	98
152	88	—	—

volumes of the spleen in an animal in which the operation was performed on the 140th day after the exteriorisation of the organ (Fig. 60). The wound cleared up excellently.

The experiments described so far have all been abdominal. Let us pass to a different type of operation, namely, excision of the superior cervical ganglion. Probably owing to my own ineptness, this has always proved a long operation, and one which has involved a good deal of manipulation to the tissues, the ganglion being deeply seated, and care being necessary to ensure the completeness of the operation; but, on the other hand, it does not involve the opening of any large cavity such as the abdomen. There is more to be expected from the results of absorption of injured tissue, and less from general sepsis, than when the abdomen is opened. The results on the volume of the spleen of the two operations are shown in Figs. 60 and 61.

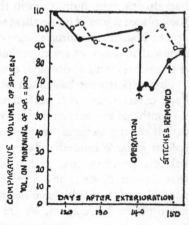

Fig. 60. Effect of opening of the abdomen on the volume of the spleen.

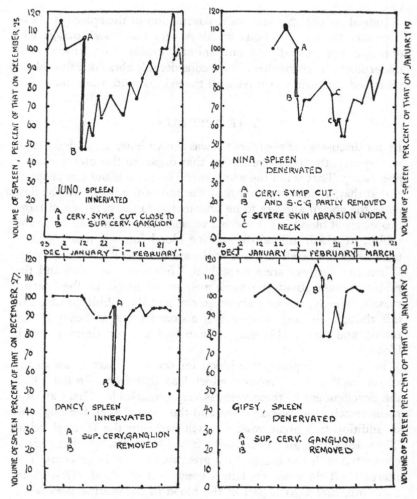

Fig. 61. Effects of operation on the volume of spleens which have and have not been denervated respectively.

Except in the duration of the after-effects, there appears to be little
to choose.

Indeed, to obtain a traumatic contraction of the spleen it is not
necessary to open the body at all. A very marked shrinkage was
obtained unexpectedly in one animal: examination proved that
underneath the collar there was a considerable abrasion; the spleen
remained somewhat shrunken till the skin wound dried up.

THE UTERUS

In our discussion of the spleen it was shown how, in the dog, blood
was apparently transferred from that organ to the uterus during
pregnancy. That is not the whole story, in fact it is but one chapter,
and at that, a chapter rather near the end of the book. In the first
place the spleen appears to be unessential to the presence of large
quantities of blood in the uterine vessels. This is well shown in the
rabbit, in which animal the uterine blood forms a much greater
percentage of the whole blood of the animal than it does in the dog.
When the foetuses are approaching "full time" the dog and the
rabbit possess about the same volume of blood in the uterine
vessels, namely, the dog about 40 c.c. and the rabbit 30 c.c., and
this though the dog may be half a dozen times as heavy as the
rabbit, and presumably may contain half a dozen times as much
blood.

In the second place, this blood for the most part is not in the
uterus itself, but in the veins which lead therefrom. In the rabbit
the development of these veins is very remarkable. There are five
main vessels; the ovarian veins and the vaginal as in the dog, and
in addition two great vessels which run into the femoral veins.
These lead from an arcade in the broad ligament, which arcade is
connected with the uterus by a great number of large veins. The
uterus itself does not contain a great deal of blood. It is clear,
therefore, that a great part of the blood in the uterine vessels has
no obvious connection with the life of the foetuses.

In the third place, we must consider the general relation of the
weight of the foetuses to that of the uterine blood in dogs, cats,
and rabbits (see Fig. 62). The impression gained from a com-

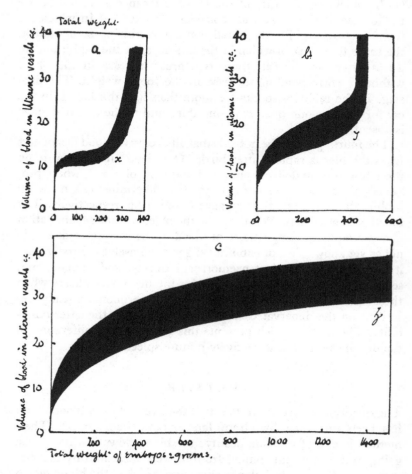

Fig. 62. Approximate relation between the quantity of blood in the uterine vessels and the total weight of the embryos. *a*, rabbits up to the 28th day; *b*, cats; *c*, dogs.

parison of these curves is that they may be divided into periods: (1) In which the quantity of blood is greatly increased, but before the foetuses have attained any considerable size. (2) In which the foetuses grow rapidly, the blood content increasing somewhat at the same time. (3) In which in the last days, in the rabbit and to a less extent in the cat, there is a great increase in the blood without a corresponding increase in the foetal weight. The final stage in the rabbit accounts for more than half the blood, in the cat for perhaps one-quarter to one-third and in the dog that stage is absent.

The fourth point to be noted is that all excessive blood disappears from the uterus rapidly after birth. One determination made on the rabbit during delivery showed but 10 c.c. of blood, where previously there had been perhaps 30; two determinations made on rabbits the day after delivery gave 6 and 9 c.c. respectively. The blood goes elsewhere. What may be the object of this accumulation and subsequent disappearance of blood in the uterine veins we do not as yet know. Two possibilities suggest themselves. Firstly, that it is an insurance against haemorrhage during confinement; and, secondly, that it provides a supply for the mammary glands when these are called into full activity. The experiments already described on the denervated spleen seem to suggest the latter possibility. The rabbit, which presents this remarkable proliferation of the uterine veins, has a relatively minute spleen.

THE LIVER

The quantitative demonstration that the liver acts as a blood depôt has been furnished by Grab, Janssen and Rein (1931). Their method is one of extreme beauty: the blood-flows in the vessels, going to the liver and coming from it, were measured by the diathermic method, and a balance-sheet was struck of the blood which entered and left the organ. When adrenaline was given the outflow from the liver far exceeded the inflow, so that a large quantity of blood, amounting in one case to over half the weight of the liver itself, was expelled into the general circulation. Table XIII shows some of Grab, Janssen and Rein's results:

Table XIII. *Blood expelled from the liver by adrenaline*

| No. | Body weight of dog kg. | Liver weight g. | Liver blood expelled | | | Adrenaline mg. |
			c.c.	Percentage of liver weight	Approximate percentage of total blood volume	
1	11·0	420	250	59	23	1/50
2	8·0	380	140	37	18	1/20
3	7·8	300	75	25	10	1/50
4	12·0	494	130	26	11	1/100
5	9·0	260	107	41	12	1/20

If the blood volume of the dog be taken as one-tenth of its body weight, the blood expelled from the liver in the above experiments amounts to 23, 18, 10, 11 and 12 per cent. of the blood volume—figures which on the whole appear to be of the same order as are capable of being expelled from the spleen.

THE SKIN

Another situation for which a claim has been put forward as a blood depôt, is the sub-papillary venous plexuses of the skin. By way of preface, let me review very briefly the arrangement of the cutaneous vessels. Any description of my own would fall far short of that given by Sir Thomas Lewis (1927): I therefore make no apology for quoting him.

In his latest account, Spalteholz describes the arteries of the cutaneous arterial network, as possessing a thick muscular coat, which diminishes relatively abruptly about the middle of the corium, where the vessels are clothed by a single layer of muscle cells. Traced further, this layer becomes imperfect, though muscle elements are still discovered on the vessels of the sub-papillary arteriolar plexus, and even on the terminal arterioles.... The venous blood returning from the papillæ passes through several networks, the first lies immediately beneath the bases of the papillæ, and receives blood from the venous limbs of the capillary loops and from the minute collecting venules formed by the union of several such capillaries. ...Almost immediately beneath this is a second venous network, the two intercommunicating freely by short venules; they are often spoken of together as forming the sub-papillary venous plexus. The blood flows deeper by numerous tributaries to the third and fourth venous networks, the former lying immediately deep to the sub-papillary arterial plexus, the latter at the level of the cutaneous plexus of arteries where the cutis and subcutis join. The superficial venous plexuses are formed by endo-thelium only according to Spalteholz; as the veins are traced more and more deeply, their walls present scattered muscle cells, which become more numerous as the last venous network is reached, and there form an imperfect sheath to the vessels. Valves and full muscular coat first appear in the subcuticular veins....

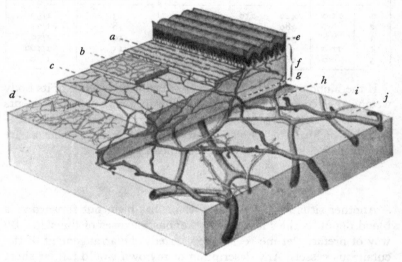

Fig. 63. Diagram of the skin and its vessels. *a*, 1st venous plexus and terminal arterioles; *b*, 2nd venous plexus; *c*, sub-papillary arteriolar plexus; *d*, 3rd venous plexus; *e*, epidermis; *f*, cutis; *g*, arched arterioles; *h*, 4th venous plexus; *i*, sweat gland layer; *j*, cutaneous arterial plexus. (After *The Blood Vessels of the Human Skin*. Sir Thomas Lewis, 1927.)

Certain deeper and direct communications have been described between arterioles and venules in the bed of the nail and pulp of the fingers and toes by Hoyer and others, and are recognised by Spalteholz....The deep communications are true arterio-venous anastomoses, the arterioles con-cerned being heavily coated with muscle.

The broad arrangement of the vessels may be followed from the accompanying diagram (Fig. 63), which is modified from that published by Spalteholz himself. It represents the manner in which the blood vessels are disposed in the palm of the hand or the sole of the foot, but the same general plan may be recognised in the skin of other parts of the body.

So much for the structure of the skin; now as to the possibility of its acting as a store. The general conception, put forward by Dr Ernst Wollheim (1927) of the Medizinische Universitätsklinik der Charité in Berlin, starts from the observation of Nicolai (1909) and Jakobj (1920), that the capillaries in the skin of the frog may be divided into two categories: the "stream capillaries", and the "net capillaries". In the stream capillaries there is a continual flow of blood; in the net capillaries varying quantities of blood can be held for a longer or shorter time side-tracked from the general circulation.

In the human foetus and at birth there are either no, or scarcely any, skin papillae. In the first months of post-natal life the capillary networks underneath the skin papillae rise from the general network of vessels. In fully developed skin, according to Wollheim, pallor corresponds to a condition in which the narrow end capillaries of the sub-papillary plexus are constricted, and only occasionally one sees a peripheral capillary loop emerging from them. Redness signifies a dilatation of the papillary capillaries, and little or no dilatation of the sub-papillary plexuses; they are in fact not visible as a complete plexus, but only as occasional vessels.

In cyanotic conditions, on the other hand, the sub-papillary networks are dilated: the bluer the cyanosis, the more the sub-papillary network comes into the picture as opposed to the papillary capillaries. Wollheim gives no data as to the amount of blood which can be lodged in the sub-papillary plexus in normal persons, but he has made numerous observations on cyanotic individuals, of which the following three may be cited (Table XIV):

Whilst sitting, the arms and legs are cyanotic, but whilst lying the cyanosis clears up, and the suggestion is that the blood which, when sitting, was stored in the sub-papillary plexuses of the limbs, finds its way into the general circulation. The measurements of the quantity of blood in circulation were made by what in England is known as the "vital red" method of Keith, Rowntree, and Geraty (1915), albeit the actual dye used was trypan blue.

The above experiments show that, coincidently with the dilatation of the vessels in the sub-papillary venous plexus, the blood volume decreases by from $\frac{1}{2}$ litre to $1\frac{1}{2}$ litres. The weak point in the experiments is, of course, that they do not exclude the possibility of other blood depôts harbouring a portion of the blood which ceases to circulate, and therefore, they only give a maximal limit in each case for the blood that can have become enclosed in the sub-papillary plexuses, not an actual measurement. Taking the lower figure mentioned above, namely, 500 c.c., we may form a

Table XIV. *Circulating blood, lying and sitting*

Case	Body weight	Volume of circulating plasma (litres)	Volume of circulating blood (litres)	
(1) Vasoneurosis	50·5	2·7	4·5	Lying: legs and arms pale; sub-papillary plexus constricted
		1·8	3·1	Sitting, one hour later: legs and arms cyanotic; intense dilatation of sub-papillary plexus
(2) Vasoneurosis with Menière's Disease	50·3	2·3	4·2	Lying: hands red, arms and face pale-red; plexus normal
		2·1	3·7	Sitting, 40 minutes later: hands and legs plum-colour; dilated plexus
(3) Scleroneurosis	66·5	2·9	5·5	Lying: arms, legs, and countenance dilated capillaries; no visible sub-papillary plexus
		2·5	5·0	Sitting, 8 hours later: legs and hands plum-colour; very dilated sub-papillary plexus

sort of estimate of the implication in the following way. The body has a surface of about two square metres, or 20,000 sq. cm.; were 500 c.c. of blood distributed over this surface as a uniform layer, that layer would be $\frac{1}{4}$ mm. in thickness.

To me the views of Wollheim have a special interest because they always carry my mind back to the circumstances which originally drew my attention to the possibility of the existence of blood depôts (Barcroft and others, 1923). On a journey to Peru, the blood volume in three subjects, as measured by the carbon monoxide method, rose when we approached the tropics, reaching a maximum as we passed through the Panama Canal, as follows:

Table XV

Place	Blood volume (litres)	Total oxygen capacity (litres)	Cabin temp. (about) (° C.)
Meakins:			
36·13 N., 43·53 W.	4·6	0·81	16
Panama	5·9	1·02	26–29
Callao	4·8	0·56	21
Barcroft:			
Cambridge	4·4	0·83	—
Panama	6·5	1·20	26–29
Callao	4·2	0·84	21
Doggart:			
Cambridge	4·6	0·96	—
Panama	6·1	1·16	26–29
Callao (near)	5·1	0·96	21

It was possible to check the hypothesis that the apparent alterations in oxygen capacity (*i.e.* haemoglobin) and blood volume were associated with the climatic temperature.

On our return to England, Dr Davies subjected himself to three days' residence in a glass chamber which was heated artificially to about 90° F. (32° C.), with the result that his blood volume rose during the residence and fell again on his return to normal (April) climatic conditions, in Cambridge, associated with east wind and, if my memory serves me, snow.

Table XVI

	Blood volume (litres)	Total oxygen capacity (litres)	Reticulated cells
Davies:			
Before (April 22nd and 23rd)	6·3±0·3	0·98±0·5	0·2
At the end of two days' residence (25th) ...	7·4±0·4	1·2±0·06	0·9
After (28th)	6·4±0·3	1·0±0·05	0·1

The immediate interest of these figures as regards the skin is that no doubt the circulation in the skin is enormously increased by exposure to heat. It is very improbable that there was any stagnation in the sub-papillary plexuses of our skins at Panama, and therefore whatever may be the storage power of the skin, that at least was probably drawn upon forthwith. But I should be far from admitting that in normal individuals under normal winter circumstances (the determinations made on Doggart and myself in Cambridge were in November) a litre of blood was stored in the sub-papillary plexuses of the skin.

THE LUNG

Recently the claim has been put forward by Max Hochrein and Keller (1932) that the lung is the important blood depôt: the claim rests on estimation of the quantity of blood which enters and leaves the lung respectively. These estimations have been made under a great number of circumstances, both by the imposition of abnormal conditions such as pneumothorax, or by the administration of drugs. Two examples will serve: (i) Short periods of inspiratory or expiratory stenosis diminish the inflow blood into the lung and at the same time increase the outflow from the organ. (ii) Administration of CO_2 increases the flow of blood into the pulmonary artery at the same time reducing the outflow. In both the instances cited the amount of blood in the lung must change materially. The claim is that the alteration is so large as to confer on the organ the status of a blood depôt. To this subject I shall return later.

THE MECHANISM OF BLOOD STORAGE

The fact that I have hitherto given no definition of a store will not have been lost upon the discerning reader. Clearly this issue cannot be postponed indefinitely, but before embarking on such a definition, let me make at least one comment on the facts which I have put forward about the skin. In the cyanosis, to which attention was directed by Wollheim, the blood lodged in the sub-papillary plexus is performing no function; in the flushing of the skin which takes place in warm climates, quite the reverse is the case. The skin is acting as an organ, not as a store; it is charged with blood because a rapid flow of blood is incidental to a great loss of heat. Blood is clearly not being stored in the skin, it is being used. Any definition of a store must recognise the distinction. Subject to the fact that the blood is not in the organ because it is being used there, we may regard a store or depôt as an organ which can, and at times does, contain a large quantity of blood capable of being diverted for use elsewhere. This is perhaps the broadest definition which can be given. It pays no heed to mechanism.

The blood in the depôt may be almost completely side-tracked. In this respect I suppose that the mechanisms of the spleen pulp and the erectile tissue of the penis may not differ greatly. In the spleen, however, the accumulation of blood is not for use in the organ, the blood is held there pending its use elsewhere: in the penis erection is a part of the functional mechanism of the organ; its erectile tissue is not a blood depôt. To go to extremes we may take an organ of quite opposite mechanism to the spleen, namely the heart. Suppose the heart gave out 80 c.c. per beat—at one time containing during systole 100 c.c. and during diastole 180, at another time during systole 0 c.c. and during diastole 80 c.c. If it were clear (1) that the alteration in mean content was wrought for the purpose of supplying the body with 100 c.c. of blood when that blood was required, and (2) that the blood served no purpose in the heart, the heart would be to the extent of 100 c.c. a blood depôt.

It is by such criteria that we must judge the claims of Hochrein and Keller. That the lung should contain more blood at some times than at others seems probable, not to say certain. One would expect such occasions to be those in which there was the maximal demand on the respiratory system. We shall see later evidence of an increased diffusion coefficient during exercise; the simple way to produce such an increase would be the exposure of a greater surface of blood.

But is there evidence that, during activity of the body, blood which has been lying in the lung is transferred to the greater circulation?

Some support for this thesis may perhaps be found in analysis

of the action of adrenaline on the minute volume recently made by
H. Barcroft (1932). The effect is complicated and depends upon
whether the drug is injected on the venous or the arterial side of
the heart, *i.e.* whether its most immediate effect is on the lesser
circulation (including the coronary vessels and the heart itself) or
the greater circulation. The complications may be simulated and
probably explained by the transfer of blood from the one circulation
to the other, but principally from the lesser to the greater. In normal
activity adrenaline is expelled from the gland into a vein: it therefore

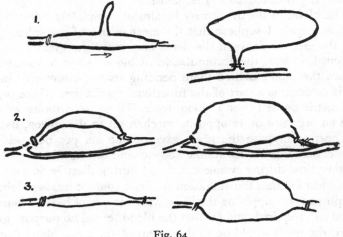

Fig. 64

reaches the periphery of the lesser circulation before that of the
greater and expels blood from the former to the latter. There are
many possible mechanisms which might serve. It is possible to
conceive such arrangements as are illustrated in Fig. 64.

In the first case the condition of the blood in the store is static,
as the avenues of entrance and of exit are the same, therefore the
blood in the store cannot be flowing. In Case 3, the blood must
be flowing, it cannot be stagnant. Case 2 is intermediate, and
according to circumstances, the blood may either be flowing or not.

These considerations would influence the length of time which
an individual corpuscle spent in the store. In Case 1, or even

Case 2, the corpuscle might be there almost indefinitely. In Case 3, which simulates that of the ordinary reservoir of a town water supply, if the filling of the skin were due to relaxation of the vessel wall on the arterial side and also on the venous, the corpuscle might go through the system more rapidly even when the store was engorged with blood than when it was empty. It therefore becomes of interest to study the length of time taken by blood to pass through any particular store. In the case of the spleen the evidence goes to show that blood may be in that organ at least half an hour. The nature of the evidence is as follows:

If carbon monoxide be given in small concentrations in the air breathed there is a lag of about half an hour in the attainment of a given percentage saturation of CO as between the blood in the general circulation and that in the spleen, the concentration in the blood of the spleen pulp being, of course, less than that in the general circulation. If, when equilibrium has become more or less established, the animal be removed to CO-free air, the lag remains, but now the blood in the spleen is richer in CO than that in the general circulation.

Some light may be thrown on the subject in the following way (Barcroft, Benatt, Greeson and Nisimaru (1931)): The skin of the right hand is rendered "blue" by exposure to cold, that of the left hand red by immersion in hot water. It is known that the blood from the skin in the latter case is practically arterial blood. The subject then suddenly inhales about 300 c.c. of carbon monoxide, the left hand still being kept hot and red, the right hand cold and blue. From time to time a prick is given to the skin of each hand, a small quantity of blood extracted, and the percentage saturation of the blood with CO is determined by the Hartridge reversion spectroscope.

Fig. 65 shows the results of such an experiment. It is clear that the blood from the cyanosed skin is a mixture, some probably coming from the papillary loops, and possibly some even from small arteries. It does not represent the blood from the sub-papillary plexus, but it represents a limiting condition. The blood in the sub-papillary plexus cannot approximate more closely to the arterial blood than does the mixed blood from the skin. The disparity in composition between the arterial blood and the sub-papillary blood cannot be less than that shown in the figure, though it may be much greater.

The percentage of CO haemoglobin in the arterial blood rises rapidly on inhalation of the gas, that in the blood from the cyanosed skin rises slowly. It is clear that there is a delay of perhaps ten minutes before a given percentage of CO haemoglobin in the corpuscles in the arterial blood is reached by the corpuscles in the sub-papillary blood, and that therefore the time of sojourn of some of these corpuscles in the sub-papillary plexus must be something

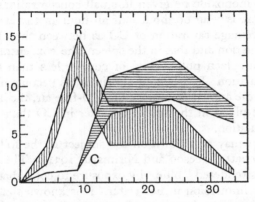

Fig. 65. Abscissa, time in min. from inhalation of CO; ordinate, percentage of CO haemoglobin; R, flushed hand; C, cyanosed hand.

measured in minutes rather than seconds, and even in minutes may run into double figures. However, the time appears to be shorter than in the spleen.

In the case of the liver, using the same sort of methods, it is not possible to show any lag between the composition of the liver blood and that of the general circulation (Fig. 66). CO sufficient to produce a considerable concentration in the blood of the anaesthetised animal is injected into the lungs through the trachea, and held there for a short time. The blood is taken for analysis (i) from the general circulation, say, the carotid or femoral artery, and (ii) from a small fresh tear made in the surface of the liver (not a cut). In only one case out of many did we observe a lag, and that was in a liver which, so far from being used as a store, was rather particularly bloodless. The liver seems to be an example of Case 3 (Fig. 64). The mechanism of blood storage in, and its discharge from, the liver have been studied by Grab, Janssen, and Rein (1931), Mautner and Pick (1915), and Dale and Poulsson (1931). If one looks back at Fig. 64, it will be apparent that if in Case 3 the store is full, there are two possible ways by which it may be emptied. The first is vaso-constriction on the arterial side, which would allow the blood to drain out into the vein without being replenished, the second is dilatation on the venous side, which allows the blood to gush out

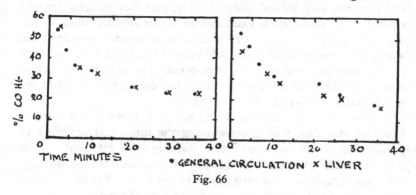

Fig. 66

faster than it enters. The two may, of course, take place simultaneously. In the case of the liver, there is no doubt about the existence of a constrictor mechanism on the arterial side: this exists both on the hepatic artery and on the arteries which supply the portal vein.

The knowledge of blood storage in the liver is difficult to trace back to a single source. The obscurity is due to the fact that it goes back to the war period, when persons in many countries, and even in laboratories in the same country, knew little of one another's work.

In 1915, Mautner and Pick, working on shock poisons, recognised a difference between herbivora and carnivora. Taking the rabbit as a sample of the former, they asserted that adrenaline causes a diminution in the volume of the liver, not by active contraction of the liver vessels, but passively, as a secondary result of contraction of the vessels of the gut. In carnivora (in which term the cat was included) Mautner and Pick reported the existence of a sensitive nervous apparatus which reacted to the presence of active proteins (as in anaphylactic shock) or protein breakdown products (as in peptone shock or histamine shock) by causing spasm of vessels in the liver. What these vessels were was not precisely decided, but they were regarded as either terminal capillaries of the portal vein, or hepatic venules or veins, or both. Adrenaline they regarded as also producing diminished flow in the way just described.

In 1920, essentially the same views were put forward by Arey and Simonds with this addition, namely, that they actually demonstrated that "histological examination of the hepatic veins in the dog reveals a relatively enormous amount of smooth muscle in the walls, thus demonstrating an adequate anatomical basis for impeding vascular flow should spasm occur. The existence of actual spasm having since been established, the correlation of structure and function in this case is complete".

It does not appear that Arey and Simonds knew of the work of Mautner and Pick, which, considering the times and the place of publication of Mautner and Pick's work, is quite intelligible. The American authors seem to have made this definite advance, namely, that they actually saw the muscles in question, and that they eliminated the capillary contraction, confining it to the veins.

The work of all the authors so far quoted makes it clear that this venous musculature is not something common to the mammalia, but that it exists in certain forms only, of which the dog is the outstanding known example. Therefore it is not an essential element in the regulation of the blood flow to the heart, though it may be an important factor in certain forms. The results already related have been elaborated and consolidated both by Pick and Mautner, and by Simonds and Brandes (1929), but it has fallen to the lot of Dale and his colleague, Poulsson, to add the finishing touch to the story, namely, that adrenaline in physiological doses dilates the sphincter of the hepatic vein, thus releasing the blood.

The claim put forward for the skin is that in the case of cyanotic individuals it contains a litre of blood which, spread over 2 sq. m.

of person, is equivalent to a layer on the average $\frac{1}{2}$ mm. thick. It is difficult to believe that so great an amount is normally stored in the skin. Let me halve that claim. And if I do that, I must out of all decency subtract something from the claim put forward by the spleen. Let us bring it to one-sixth of the blood volume, while that of the liver comes to a fifth. We would then have

> Blood in store in the liver   ...   20 per cent.
> „   „   „   spleen ...   16   „
> „   „   „   skin   ...   10   „
>   ——
>   46   „

of the total blood volume.

It is not impossible that if all the stores are filled simultaneously nearly half of the blood in the body may be withdrawn from circulation. But we know almost nothing about whether the stores can be filled to their full extent simultaneously, or of the order of their filling. One solitary fact points a little in that direction, namely, that the two largest—the liver and the spleen—are both emptied by administration of the same substance—adrenaline. Possibly the lung participates in this also.

We seem to have drifted far into the region of speculation; the immediate point, however, is substantiated, namely, that there is nothing inherently absurd about the conception that a litre more of blood may be in circulation at one time than at another. The mechanism of this unloading of blood from the liver was, according to Rein, principally a sudden constriction of the vessels leading thereto; so that the liver, turgid with blood, emptied itself into the hepatic vein without the replacement of the blood from the arterial end. This is not, however, the whole story, because several authors have shown that besides this mechanism, which I suppose is present in all mammals, there is another at the roots of the hepatic vein which Dale and his colleagues have termed the venous sluice. The venous sluice, as I am sure Bauer, Dale, Poulsson and Richards (1932) would agree, can only be of subsidiary importance, because it is only present in occasional forms, e.g. the dog, but many vertebrate forms exist in which no evidence of a venous sluice has been adduced.

REFERENCES

AREY, L. B. and SIMONDS, J. P. (1920). *Anat. Record.* **18**, 219.
BARCROFT, H. (1932). *J. Physiol.* **76**, 339.
BARCROFT, J. (1925). *The Lancet*, **1**, 319.
—— (1930). *J. Physiol.* **68**, 375.
BARCROFT, J. and BARCROFT, H. (1923). *Ibid.* **58**, 138.
BARCROFT, J., BENATT, A., GREESON, C. E. and NISIMARU, Y. (1931). *Ibid.* **73**, 344.
BARCROFT, J., BINGER, C. A., BOCK, A. V., DOGGART, A. H., FORBES, H. S., HARROP, G., MEAKINS, J. C. and REDFIELD, A. C. (1923). *Phil. Trans. Roy. Soc.* B, **211**, 351.
BARCROFT, J. and FLOREY, H. (1929). *J. Physiol.* **68**, 181.
BARCROFT, J., HARRIS, H. A., ORAHOVATS, O. and WEISS, R. (1925). *Ibid.* **60**, 433.
BARCROFT, J. and ROTHSCHILD, P. (1932). *Ibid.* **76**, 447.
BARCROFT, J. and STEPHENS, J. G. (1928). *Ibid.* **66**, 32.
BAUER, W., DALE, H. H., POULSSON, L. T. and RICHARDS, D. W. (1932). *Ibid.* **74**, 343.
BENHAMOU, E., JUDE, MARCHIONI and NOUCHY (1929). *C.R. Soc. Biol.* **100**, 456, 458, 461.
BINET, L., CARDOT, H. and FOURNIER, B. (1927). *Ibid.* **96**, 521.
BINET, L., CARDOT, H. and WILLIAMSON, R. (1926). *Ibid.* **95**, 262.
BINET, L. and FOURNIER, B. (1926). *Ibid.* **95**, 1141.
CASTEX, M. R. and DI CIO, A. V. (1929). *La Prensa Médica Argentina.*
CRUICKSHANK, E. W. H. (1926). *J. Physiol.* **61**, 455.
DE BOER, S. and CARROLL, D. C. (1924). *Ibid.* **59**, 381.
FELDBERG, W. and LEWIN, H. (1928). *Pflügers Arch.* **219**, 246.
GRAB, W., JANSSEN, S. and REIN, H. (1931). *Klin. Woch.* **33**, 1539.
GRAY, H. (1854). *On the Structure and use of the Spleen*, p. 341. London.
HANAK, A. and HARKAVY, J. (1924). *J. Physiol.* **59**, 121.
HARGIS, E. H. and MANN, F. C. (1925). *Amer. J. Physiol.* **75**, 180.
HEGER, P. (1894). *Upon Unequal Diffusion of Poisons into the Organs*, p. 5. Oxford.
HOCHREIN, M. and KELLER, C. J. (1932). *Arch. Exp. Path. Pharm.* **164**, 529, 552.
HODGKIN, T. (1822). *Edin. Med. Surg. J.* **18**, 83.
IZQUIERDO, J. J. and CANNON, W. B. (1928). *Amer. J. Physiol.* **84**, 545.
JAKOBJ, W. (1920). *Arch. Path. Pharm.* **86**, 49.
KEITH, N. M., ROWNTREE, L. G. and GERATY, J. T. (1915). *Arch. Int. Med.* 547.
LEWIS, Sir THOMAS (1927). *The Blood Vessels of the Human Skin and their Responses*, p. 1. London.
MCROBERT, G. R. (1928). *Ind. J. Med.* **16**, 553.

MAUTNER, H. and PICK, E. P. (1915). *Münch. Klin. Woch.* **34**, 1141.
NICOLAI, G. F. (1909). Nagel's *Handbuch der Physiologie des Menchen.* Braunschweig.
SCHEUNERT, A. and KRZYWANEK, F. W. (1926). *Pflügers Arch.* **212**, 477.
SIMONDS, J. P. and BRANDES, W. W. (1929). *J. Pharm. Exp. Therap.* **35**, 165.
WOLLHEIM, E. (1927). *Klin. Woch.* **45**, 2134.

CHAPTER VII

EVERY ADAPTATION IS AN INTEGRATION

FOETAL RESPIRATION

The physical and chemical properties of the blood even in man are not constant. In most respects they vary to an extent which is large enough to be measured with reasonable accuracy. Change of external environment alters them. The most that can be asked is, firstly, that they should alter in an orderly way, secondly, that there should be such a degree of stability about these alterations that when the organism returns to its normal environment, the *internal milieu* should regain its original properties, and thirdly that the alteration in any one property should be as little as possible. It is the last proposition which I wish to discuss.

A great result may be obtained by the summation of several small changes, and a still greater result may be obtained by their multiplication. The amount of blood driven round the body is susceptible of a variation in the region of ninefold. That is not true either of the pulse rate or the stroke volume; it is achieved by variation in each of the order of threefold and by the multiplication of the two.

In the following chapters (VII–IX) I propose to cite some circumstances to which the body must needs adapt itself, and to give an analysis of the form which that adaptation takes, showing how rather large effects may result from the cumulative action of a number of factors, each of which alters in a lesser degree.

The first condition which I shall discuss is that of pregnancy. The mother adapts herself to the fact that within her a large organism is developing rapidly; on the one hand, that development takes place at her expense; on the other hand, its success depends upon the maintenance of her health. We will concern ourselves here with the transference of oxygen from the mother to the foetus. As the foetus requires ever increasing quantities of oxygen, the mother must adapt herself to the provision of that gas. She may adopt means either mechanical or chemical, or both, which will aid the transition of oxygen from her blood to that of the embryo; but these alterations must not be so far-reaching as to be a serious menace to her own condition, and naturally the more highly organised she is, the less will be the possible alteration which she can sustain consistent with safety to her own health. A certain quantity of oxygen is necessary for the foetus; this oxygen is carried to the placenta in the haemoglobin of the maternal blood. Into that organ also goes one branch of the vascular system of the foetus. The two blood streams, the maternal and the foetal, though they do not mix, come into such close contact in the placenta that the oxygen from the maternal blood can pass through the vessel walls (or wall) by which the two streams are separated. There appears to be no doubt that the passage is due to diffusion. What are the conditions under which this diffusion takes place, and how are they modified as the requirements of the foetus increase? The first factor for consideration is the quantity of maternal blood which reaches the placenta.

Clearly if the maternal blood must impart a certain number of cubic centimetres of oxygen to the foetus, the greater the volume of maternal blood involved the less will it be reduced, consequently the higher will be the average pressure of oxygen in the maternal blood sinuses and the better the opportunities for diffusion. The mother, however, cannot improvise unlimited quantities of blood either by manufacturing it or transferring it from elsewhere. The quantity of blood in the uterus and its vessels has been investigated by Paul Rothschild and myself (1932) in the cat and dog and in greatest detail in the rabbit. Further work on the subject has been done by Herkel, Hill and myself (1933).

Table XVII gives the maximal normal quantity of blood present in the uterine vessels during pregnancy, for three species.

Table XVII

	Rabbit	Cat	Dog
Quantity of blood (c.c.) in uterine vessels (approximate):			
Not pregnant	2	1·5	1–2
Maximum during pregnancy	32	37	41
Approximate weight of blood per gm. of foetus	0·13	0·07	0·03

The absolute quantities are surprisingly alike in spite of the difference in weight of the mothers, the dogs being mostly about 10–15 kg. while the rabbits and the cats were 2–3 kg.

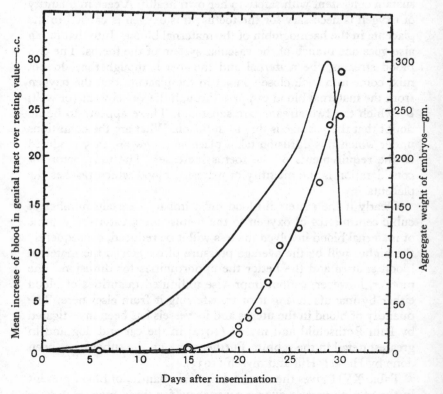

Fig. 67. Circles show the weights of the embryos; the line without circles, the increment of blood.

The volume of maternal blood which traverses the uterus per minute is of the same order as the volume of blood in the vessels, therefore we may say that, for the rabbit, 30 c.c. of blood per minute, containing perhaps 5 c.c. of oxygen, is the contribution which the mother can make.

One important point is that the great expansion of the circulation

takes place before the embryos are of more than negligible bulk
(Fig. 67). The stage is set before the play commences. From this
stream of blood almost no oxygen is taken when the embryos are
very small; as they grow, the degree of reduction of the maternal
blood increases as shown in Fig. 68.

The rabbit as shown by Mossmann (1926) achieves the transference of oxygen from the maternal to the foetal blood by a very
remarkable arrangement of vessels. These are disposed so that the

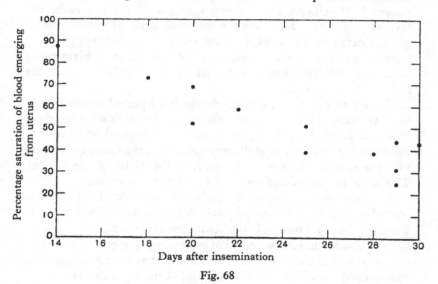

Fig. 68

maternal and foetal vessels are parallel as well as being in juxtaposition; the dark blood enters from the foetus in proximity to the
maternal venous blood leaving the placenta, whilst the blood
leaving the placenta for the foetus is in proximity to the maternal
arterial blood. The foetal blood, therefore, can pick up oxygen
along the whole course of the vessel and leave the placenta only
slightly less red than that entering from the mother.

The above arrangement is an unusual one, and in other animals
the question arises: is there any chemical mechanism to facilitate
the diffusion of oxygen?

Previous workers in this field had been Huggett (1927), and also Haselhorst and Stromberger (1930, 1931 *a* and *b*), both of whom had worked by modern methods. The latter workers had performed an elegant research on human subjects who had been objects of Caesarian section; Huggett worked on goats.

Both the above authors made observations on the oxygen dissociation curve and their results differed, for while Huggett described the curve of the foetal blood as being to the right of the maternal, Haselhorst and Stromberger placed the curves in the reverse positions. The latter result has been obtained but in greater detail by Eastman, Geiling and De Lawder (1933). They found the oxygen dissociation curve of the foetus at birth to be in the normal position whilst that of the mother was to the right of the normal.

There was, however, no research which purported systematically to have followed the possible alterations throughout the whole course of pregnancy. Huggett's goats were operated on as near to full term as possible, whilst the experiments of the German workers took place, in the nature of the case, at the birth of the infants. The same is true of the work of the American authors.

A detailed description of the technical methods which we* employed would, here, be out of place. A convenient animal to study is the goat, and we were able to obtain a series of goats, all "stocked" about the same time. The period of gestation in the goat is 21 weeks, the weight of the kid at birth being somewhat over 2000 gm. We commenced our observations at the end of the seventh week, when the foetus weighs about 10 gm., and operated on one goat weekly. Of the series, two animals proved not to be pregnant. A third, unfortunately the last, was found to have congestion of the lungs; while therefore it might have formed the starting-point of a very interesting research on the effect of acute pulmonary trouble in the mother on foetal circulation and respiration, it could yield no results of value as a final experiment in our series of normals. Between the seventh and the eighteenth weeks, however, we have a tolerable series.

* Preliminary communication: Barcroft, 1933, Nov. 4, p. 1021.

The Dissociation Curve of the Mother

This may be considered first (Fig. 69). The curves of control goats which had not been stocked within 21 weeks of the date of

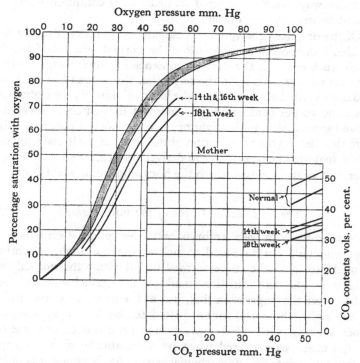

Fig. 69. Blood of goat. Left hand: oxygen dissociation curves at 50 mm. CO_2 pressure. Dotted area = position of curves of non-pregnant animal; lines = curves at stated periods of pregnancy. Right hand: portions of CO_2 dissociation curves.

investigation are not quite identical, but fall within a very restricted area. This is true whether the blood be "whipped" or whether it be diluted with a measured quantity of oxalate-fluoride mixture—the oxalate to prevent clotting, the fluoride to prevent glycolysis. Anaesthetised non-pregnant goats yield curves also within this area, though if urethane is used the curves tend to be at the right-hand edge.

The dissociation curves of the goats of our stocked series which were not pregnant fell also within this area. They were operated on in every way like the pregnant animals, their condition being revealed on opening the uterus.

Of the pregnant goats, that investigated in the seventh week also yielded a dissociation curve within the normal area, but from the tenth week onwards the curve commenced to drift to the right of the normal area, that is, in the sense that the oxygen became more and more easily detachable from the haemoglobin. By the eighteenth week the 50 per cent. saturation point of the maternal curve had moved some 9 mm. to the right of the mean normal position. Note here that the degree of inflection of the curve is unaltered; any one of the maternal curves might be derived from any other by a mere alteration of the scale on which the horizontal component is drawn.

The Dissociation Curve of the Foetal Blood

Ninth and tenth week embryos were the youngest on which we experimented. On the blood of each we made a single determination. Here again the curve seemed to fall within the normal area (Fig. 70). The determinations made on the weeks just subsequent at first gave the impression that, just as the maternal curve drifted to the right, the foetal curve drifted to the left—a symmetrical process taking place. That impression was not altogether borne out by the facts—as appeared when we were able to obtain a larger number of points on any particular curve. At all events it is only a partial representation of them. It is true that in the fifteenth week the 50 per cent. saturation point of the foetal blood is to the left of that in the tenth or eleventh week : but a new factor appeared. The foetal dissociation curve was at this stage of a different shape from that of the mother. It was less inflected : at about 90 mm. the two cross, and from this point downwards they diverge. The proportionate distance between them, expressed as a function of the distance of one or other from the ordinate, increases progressively as the saturation diminishes. Actually between 80 and 40 per cent. saturation the foetal and maternal curves are roughly parallel separated in the fourteenth week by about 13 mm.

At a later stage the gap is greater, the dissociation curve of the mother having shifted further to the right, whilst that of the foetal blood remains in much the same place as in the fourteenth week,

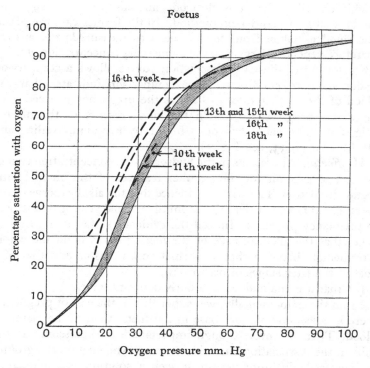

Fig. 70. Blood of foetal goats; oxygen dissociation curves at 50 mm. CO_2 pressure. (Dotted area as in Fig. 69.)

but towards the end of pregnancy the degree of inflection seems to vary within wider limits, approximating in some cases to that of the maternal blood.

Attempt to explain the Movement of the
Maternal and Foetal Curves

I. *Maternal.* The alteration in the position of the dissociation curve appears to be a matter of hydrogen-ion concentration.

Experiments have been made on this subject by Havard and Dickinson using the glass electrode, and also by Keys (1934) by the Hasselbalch-Henderson method. In the former case the pH of the blood as taken from the vessels was measured; the answer therefore depended largely on the momentary concentration of CO_2. The plotting of the CO_2 dissociation curves allows a comparison under standard conditions. This comparison indicates that the blood of the mother loses in base as the pregnancy proceeds, for there is a progressive drop in the alkali reserve of the plasma (Fig. 69). The same has been found by Eastman, Geiling and De Lawder (1933, see Fig. 2).

II. *Foetal.* Alterations in the hydrogen-ion concentration cannot be said to account for the peculiar features of the blood in the foetal circulation. There is no evidence that the alkali reserve of the foetal blood is higher than that from the normal goat or that it rises as pregnancy proceeds. Indeed the evidence so far as it goes is:

(1) that the alkali reserve of the foetal plasma is on the whole intermediate between that of normal goat's plasma and that of a goat in the later stages of pregnancy;

(2) that it is maintained at a fairly constant level.

From these, incidentally, we gather that at the end of pregnancy there is considerably more base in the foetal than in the maternal blood. That, however, though important is a digression. If the shift in the dissociation curve is not to be explained on the ground of increased alkalinity, we must seek a solution in some other direction.

It is natural to ask whether the fundamental characters of the foetal and maternal haemoglobins are identical. In this connection let us go back to the demonstrated abnormality in the inflections of the dissociation curve of the foetal blood. The comparison between the maternal and foetal haemoglobin has been taken up separately by both McCarthy (1933) and Hall (1934), and along different lines.

McCarthy prepared the purified haemoglobin from the blood by Adair's method: he obtained solutions of the same order of concentration as blood and in considerable quantities. He then determined the dissociation curves by the van Slyke technique at a pH of 6·8.

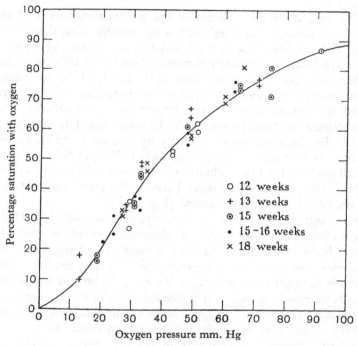

Fig. 71. Oxygen dissociation curves of haemoglobin prepared in concentrated solution from goats at stated periods of pregnancy; pH 6·8 (McCarthy).

All the solutions prepared from the maternal blood gave substantially the same dissociation under standard conditions (Fig. 71). Not only so, but the curve of the mother's haemoglobin did not differ from that of a normal goat, which is confirmation of a negative kind of the point already made, namely that the maternal drift is due to alteration of hydrogen-ion concentration.

When, however, we come to the foetal haemoglobin, there is no such identity between one goat and another. The series extends from the thirteenth week onwards. The foetal haemoglobins, however, have this in common: the curves are all to the left of the maternal, showing that the haemoglobin of the foetus has a higher affinity for oxygen than that of the mother. An example is given in Fig. 72 *a*.

The general similarity between the dissociation curves of the foetal and maternal haemoglobins is comparable with that of the foetal and maternal bloods from which those haemoglobins were made. Indeed, the similarity is greater than might be anticipated, for the hydrogen-ion concentration of the inside of the corpuscles is not known; the blood was equilibrated at 50 mm. CO_2 pressure.

Hall worked on dilute solutions by spectroscopic methods, using no elaborate methods to purify the haemoglobin, but merely reducing the concentration of impurity by putting a small quantity of blood or haemoglobin into water adjusted to pH 6·8.

Hall's results tend throughout to be further to the left than those of McCarthy, but this difference between the foetal and maternal haemoglobins is as clearly marked (Fig. 72 *b*).

It is interesting to note, by the way, that according to Hall's observations the "same sort" of difference existed between the haemoglobin of the chick (in the egg) and that of the hen. This difference seems reasonable enough, since the function of haemoglobin is to acquire oxygen where the gas is to be had and subsequently to give it up where it is needed. In the mature organism the pressure at which the oxygen can be acquired is high, while the activity of the organ demands that the oxygen should be detached with ease. Foetal respiration is of the opposite character, the average pressure of oxygen in the placental capillaries is low, whilst the tissues supplied are leading a vegetative existence.

Indeed the conception of there being more than one sort of haemoglobin even in ordinary blood is not revolutionary; it may be that the difference between foetal and maternal haemoglobins lies in the proportions in which the two are mixed. Reichert and Brown (1909) pointed out that there were in blood two forms of haemoglobin. Geiger (1931), by a process of electro-dialysis, separated from blood two haemoglobins which had dissociation

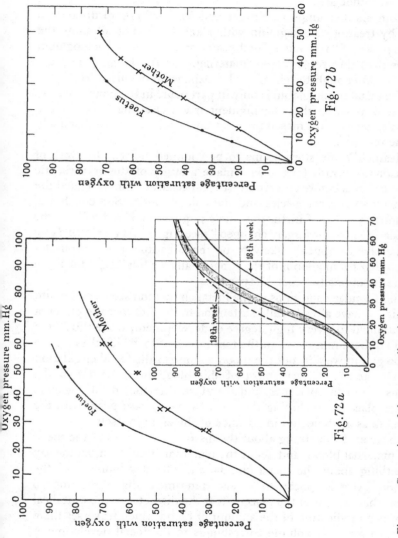

Fig. 72a. Comparison of the oxygen dissociation curves of maternal and foetal haemoglobin at pH 6·8 and of the maternal and foetal blood, at 50 mm. CO_2 pressure, from which the haemoglobin was prepared (McCarthy).

Fig. 72b. Comparison of oxygen dissociation curves of foetal and maternal haemoglobin in dilute solution, pH 6·8 (Hall).

curves of different degrees of inflection, but it is not certain that they were not artifacts.

From another angle v. Krüger and Bischoff (1925) discovered that by treating haemoglobin with alkali it did not all show the same power of "resistance" to the same and that foetal haemoglobin differed in that respect from maternal. Dr Brinkman (1934) has been good enough to tell me of his expansion of this work.

The foetal haemoglobin is only in part made in the bone marrow, and it is possible that the pigment made elsewhere, e.g. in the vascular cords of the mesoderm, may not be identical with that made in the marrow.

Meanwhile the story seems to be rounded off by the work of Christianna Smith (1932) who made a study of the erythrocytic picture in rats before and after birth; especially she investigated the diameter of the corpuscles and the colour index. She concludes, "About the time of birth, the whole population of red cells in the circulating blood is largely changed" (see Fig. 73). The raison d'être of the change appears likely to be the substitution of corpuscles containing haemoglobin of the functionally "adult" type for those of the functionally "foetal" type.

The vascular conditions in the foetal circulation seem to be quite special. I have already called attention to the fact that the placenta has attained to a very high degree of development, while the foetus is still quite diminutive. In the tenth week of foetal life the embryo is 100 gm. in weight, but the blood volume of the foetal circulation may be 40 c.c. Clearly most of this blood must be not in the foetus but in the placenta, and therefore that the foetal elements in the placenta (including the blood) develop pari passu with the placenta as a whole and in advance of the embryo itself.

I have said something about the dissociation curves of the foetal and maternal blood, and about the maternal circulation, let me say something about the foetal circulation. The development of the umbilical vessels appears to be related in time to that of the embryo rather than to that of the placenta. If the diameter of the umbilical veins is an indication of the amount of blood which traverses them in unit time, that volume corresponds to the foetal development. Hence the picture at the end of the tenth week is that of a large

placenta, 40 c.c. of blood shared between the placenta and the
foetus but principally in the placenta, and circulating very slowly.
Although this blood is poor in haemoglobin, yet its volume is so
considerable as to render the actual quantity of pigment out of all

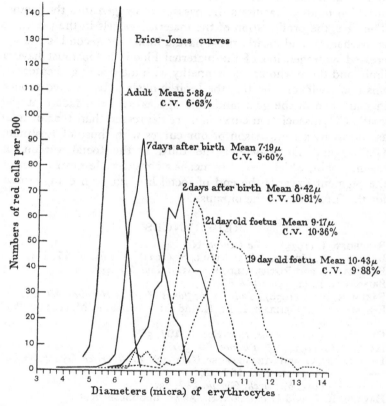

Fig. 73. Showing population of erythrocytes of stated diameters before and after
birth in the rat (Christianna Smith).

proportion to the foetus. The ratio of the total oxygen capacity of
the blood in the foetal circulation at this stage to the total foetal
volume is about 4 to 100, whilst at the end of the eighteenth week
it is only 1 to 100. It is well known that, towards the end of foetal

life, much of the haemoglobin is broken down and the iron is stored in the liver to be rebuilt once more into blood pigment post-natally.

To summarise. Apart from the alterations in the foetal circulation about which we know little, there are three principal alterations each of which tends to facilitate the passage of oxygen into the foetus. The first, the proliferation of the maternal vessels in the placenta, is mechanical and clearly must have a limit; the second is the increased hydrogen ion of the maternal blood; it too must have a limit, and those who are in sympathy with the opening chapters of this book will consider that the limit is likely to be reached sooner in man than in the goat, and this agrees with the fact that the goat's CO_2 dissociation curve is more depressed than that of man as shown by a comparison of our curves with those of Eastman, Geiling and De Lawder; the third is the foetal variant of haemoglobin, which probably means a considerable destruction of the pigment towards the end of foetal life, the iron being stored for the future use of the organism.

REFERENCES

BARCROFT, J. (1933). *The Lancet*, Nov. 4, 1021.
BARCROFT, J., HERKEL, W. and HILL, S. (1933). *J. Physiol.* **77**, 194.
BARCROFT, J. and ROTHSCHILD, P. (1932). *Ibid.* **76**, 447.
BRINKMAN, R. (1934). *Ibid.* **80**, 377.
EASTMAN, N. J. (1930). *Bull. of the Johns Hopkins Hospital*, **47**, 221.
EASTMAN, N. J., GEILING, E. M. K. and DE LAWDER, A. M. (1933). *Ibid.* **53**, 246.
GEIGER, A. (1931). *Proc. Roy. Soc.* B, **107**, 368.
HALL, F. G. (1934). *J. Physiol.* **80**, 502.
HASELHORST, G. and STROMBERGER, K. (1930). *Z. Geburtsch. Gynäk.* **98**, 49.
—— —— (1931 *a*). *Ibid.* **100**, 48.
—— —— (1931 *b*). *Arch. Gynäk.* **147**, 65.
HAVARD, R. E. and DICKINSON, S. As yet unpublished.
HUGGETT, A. ST G. (1927). *J. Physiol.* **62**, 373.
KEYS, A. (1934). *Ibid.* **80**, 491.
McCARTHY, E. F. (1933). *Ibid.* **80**, 206.
MOSSMANN, H. W. (1926). *Amer. J. Anat.* **37**, 433.
REICHERT, E. T. and BROWN, A. P. (1909). *The Crystallography of Haemoglobins.* Carnegie Inst. Washington Pub.
SMITH, C. (1932). *J. Path. Bact.* **35**, 717.
v. KRÜGER, F. and BISCHOFF, H. (1925). *Ber. ges. Physiol.* **32**, 696.

CHAPTER VIII

EVERY ADAPTATION IS AN INTEGRATION
(continued)

ADAPTATION TO EXERCISE

Let me commence by drawing your attention to four sets of measurements which will be considered at first quite separately, but which later will prove not to be so unconnected as might appear at first sight.

The first which I shall select is the quantity of haemoglobin in each cubic centimetre of blood. This quantity has been of great interest to pathologists for long enough because of the very wide limits within which it varies—ranging through all stages of anaemia, when it is diminished, and of polycythaemia when it is increased. In what units is it to be measured? The older workers judged of the haemoglobin content of the blood either by the colour or the specific gravity of that fluid. Let us drop the specific gravity and concentrate on the colour. All sorts of scales were invented, the general nature of which was as follows: A standard was made of some suitable red material—Gowers used coloured jelly, Fleischl coloured glass, etc.—which just matched the colour of normal blood diluted precisely 100 times. The colour value or haemoglobin value of normal blood was then 100; if a given blood could only stand diluting 50 times before it matched the standard, that blood had a haemoglobin value of 50, and so on.

Haldane (1901) showed that the haemoglobin value on Gowers' empirical scale varied directly with the oxygen carrying power or "oxygen capacity" of the blood. He concluded that the haemoglobin of blood with a value of 100 had an oxygen carrying power of 18·5 c.c. of O_2 at N.T.P. for every 100 c.c. of blood. That concentration of haemoglobin diluted 100 times is the standard for most haemoglobinometers used in Great Britain. Within recent years American observers have placed the normal haemoglobin value higher than Haldane did. It may be that the blood of the normal man in America contains more haemoglobin than that of his "opposite number" in Britain, but the work of Jenkins and Don (1933) suggests that the haemoglobin concentration is the same and that the van Slyke pump extracts more oxygen from human blood than does potassium ferricyanide which was largely used by Haldane, but, whatever the explanation, I will subscribe to the proposition that the direct measurement of oxygen is theoretically superior to the use of coloured standards—and that the oxygen carrying power furnishes a more satisfactory standard of units for the expression of haemoglobin content than does an arbitrary standard. Let us state quite frankly then, that, as far as is known, the normal oxygen carrying power of blood is 20 c.c. of oxygen for every 100 c.c. of blood.

When we concern ourselves with the units used for the expression of the quantity of oxygen in blood unsaturated with that gas, we find ourselves in greater difficulties. The simplest form of expression is a mere statement in absolute units, for instance, that 100 c.c. of a given sample of venous blood contains 10·5 c.c. of oxygen. For many purposes, however, it is of more interest to know the relative proportions of oxy- and reduced haemoglobin present, and to have units expressing the degree of oxidation rather than the actual quantity of oxygen in the blood. We speak of blood as being 50 per cent. saturated. Clearly if we know the oxygen capacity it is easy to convert the one set of units into the other. Thus, if a sample of blood has an oxygen capacity of 20 volumes per cent. and be 50 per cent. saturated, the blood contains 10 volumes per cent. of oxygen. Or again, if we consider the blood as it circulates around the body, we can regard the amount of oxygen taken out by the tissues as being either a percentage of the whole oxygen carried, *i.e.* of the oxygen capacity, or as an absolute quantity. Regarded as an absolute quantity it is called the *absolute oxygen transport*—as compared with the *percentage oxygen transport*.

We have therefore a general relation,

$$A.O.T. = O.C. \times \frac{O.T. \text{ per cent.}}{100},$$

$$O.C. = \frac{A.O.T.}{O.T. \text{ per cent.}} \times 100.$$

It is of course possible to express the above result graphically, the oxygen capacity appearing as a straight line, while the absolute oxygen transport and the percentage oxygen transport are the abscissa and the ordinate respectively.

In the following argument we shall have to consider a number of factors, some of which are currently spoken of in cubic centimetres, some in litres, some in millimetres, some in percentages

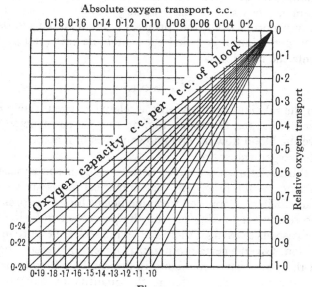

Fig. 74

and so forth. For the sake of simplicity of treatment I propose to state them all in centimetres. Instead of the percentage oxygen transport let us have the *relative* oxygen transport (R.O.T.) for 1 c.c. of blood and not for 100:

$$\text{O.C.} = \frac{\text{A.O.T.}}{\text{R.O.T.}}.$$

R.O.T. is the proportion of the whole oxygen capacity of 1 c.c. of blood which is reduced in the tissues or oxidised in the lung (Fig. 74).

The second relationship of which I wish to speak is connected with the rate of blood flow. There are many methods of measuring the quantity of blood which passes round the circulation in a minute. In animals this can be done directly with stromuhrs, etc.; in man some indirect method must be sought. This indirect method (with which the name of Fick is associated) consists in measuring (1) the oxygen taken in per minute, (2) the oxygen taken up by each cubic centimetre of blood as it traverses the lung. The latter is the measurement to which I have already alluded as the absolute oxygen transport (A.O.T.); the former, *i.e.* the total oxygen used by the body per minute, is called the metabolic rate (M.R.). If B.F. represents the blood flow:

$$\text{B.F.} = \frac{\text{M.R.}}{\text{A.O.T.}}.$$

As in the relationship between the relative oxygen transport and the absolute oxygen transport, Fig. 75 may be constructed which will give the blood flow for any known values of metabolic rate and absolute oxygen transport.

The third relationship which I wish to study is some measure of the efficiency of the lung. In what numerical terms are we to describe an efficient as opposed to an inefficient lung? At present we have no reason for looking beyond the process of diffusion as the cause of gaseous exchange; therefore let me seek a measure in terms of diffusion. Given a certain difference of oxygen pressure between that in the pulmonary alveoli and in the capillary blood, the more efficient the lung, the greater is the quantity of oxygen which passes through its wall. Numerically, if M.R. (metabolic rate in terms of oxygen) be the quantity of oxygen absorbed per minute, A.P. the partial pressure of oxygen in the aveolar air, and C.P. the average partial pressure of oxygen along the length of the capillary:

$$\text{M.R.} = \text{K} \times (\text{A.P.} - \text{C.P.})$$

(K is called by Krogh the diffusion coefficient), or

$$\text{K} = \frac{\text{M.R.}}{\text{A.P.} - \text{C.P.}}.$$

The details of the way in which K is measured need not detain us, they are given in Krogh's (1915) paper. As applied to myself on one occasion κ came out to be 36 in terms of the units used by Krogh, *i.e.* for each mm. of difference of pressure between the oxygen in the alveolar air and that in the blood, 36 c.c. of oxygen per minute traversed the lung wall. The lung of another man

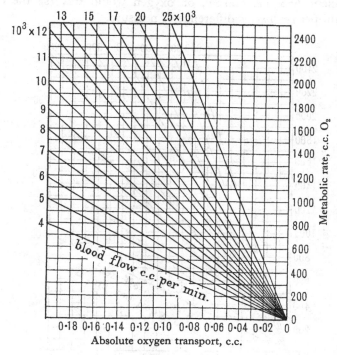

Fig. 75

whose coefficient was measured at the same time gave a value of 46. His was therefore a more efficient lung, either because his lung presented a greater area of capillaries for the reception of the oxygen, or because the capillary wall was thinner, or for both reasons. Nor is it surprising that he was possessed of a better pair of lungs for he was a very fine swimmer. Brought up on the banks of Lake Ontario he told me that when he was a boy, if he wished to visit a friend 5 miles away, he was as likely to swim the distance as to walk it. The most efficient lung which I have come across was that of a man, Rogers, whom we met in Peru. His lung had a diffusion coefficient of about 60—nearly twice as good as my own. Here again let us keep to centimetres and call his diffusion coefficient 600, *i.e.* 600 c.c. of oxygen would traverse the epithelium per *centimetre* difference of pressure.

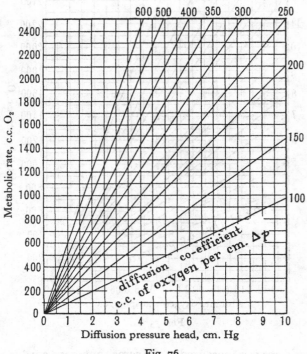

Fig. 76

The relations which I have discussed may be expressed graphically (Fig. 76). If the oxygen used (the metabolic rate) be made the ordinate and the difference between the average pressures in the alveolar air and the capillary blood be made the abscissa (*i.e.* the diffusion pressure head) the various values of the diffusion coefficient appear a series of straight lines, according to the general relation

$$\text{D.C.} = \frac{\text{M.R.}}{\text{D.P.H.}}.$$

There is yet one more relationship which falls into the same sort of form as those which we have discussed, but which requires slightly more complicated treatment in order to get it into that form. The efficiency of the lung depends upon two things: (1) the mechanical configuration of the organ, regarded as a medium for diffusion—that combination of the area of the capillary bed, the thinness of the capillary wall, and the rate at which oxygen can diffuse through it which we have called the diffusion coefficient, and (2) the quantity of blood or rather the quantity of haemoglobin in the blood, which circulates through the vessels and carries the oxygen away. These two factors must clearly be adjusted the one to the other. It would be useless circulating large quantities of blood through a lung, the walls of which were impermeable to oxygen, or having a lung which was mechanically perfect as a medium for diffusion, but through which there was a deficienc circulation. We may therefore ask—what is this relation? For each unit in the diffusion coefficient, how much haemoglobin is driven through the lung? In other words, what is the relation of $\frac{\text{H.F.}}{\text{D.C.}}$ which may be called the specific haemoglobin flow. The relation of the haemoglobin flow to the diffusion constant represents at first sight something more obscure than some of the other relationships which we have discussed. Yet it is very important, and in illustration of its importance I may be excused a homely illustration which is coupled with it in my mind.

Imagine a crowd say of 20,000 or 30,000 persons from a football match arriving at the station adjacent to the ground and being taken away by train. There are two critical points in the smooth running of this operation—the railway station and the supply of trains. The crowd cannot reach the platform faster than the barriers will allow the people in; when they have reached the platform, unless there are sufficient trains the passengers will accumulate and ultimately prevent the further influx of the crowd. On the other hand, it would be useless to have too many or too long trains for, unless the number of barriers was correspondingly increased, the trains would leave half empty. Crudely then, the number of barriers corresponds to the diffusion coefficient of the lung, and the number of railway carriage seats which pass through the station per minute represents the haemoglobin flow. The number of people who can get through the barriers depends (1) on the barriers, and (2) on the relative crush inside and out, whilst the crush inside in its turn depends upon the rate at which the trains take the throng away.

Without pushing the analogy too far, it is sufficiently evident that the ratio of trains to barriers is of paramount importance, whilst the absolute adequate number of each depends upon the size of the crowds to be dealt with. It is perhaps worth remembering that the system will not break down if either the barrier accommodation or the train accommodation is excessive—it merely becomes wasteful.

With these limits in our minds, let us turn to the lung. Suppose, as appears to be the case in persons who have suffered from gas poisoning, the diffusion coefficient becomes reduced so that the oxygen comes through much more slowly than normal, for a given pressure head. What will be the repercussion of the alteration in the lung on the whole system? Granting that the amount of oxygen needed is small, it can traverse the lung by increasing the pressure head, *i.e.* by lowering the oxygen pressure in blood, but this process can only help as a makeshift, for there must come a time when lowering the oxygen pressure in the blood means one or both of two things: (1) the blood is not carrying a full load of oxygen, or (2) it cannot impart it to the tissues. But, in any case, the man is using up during rest the vascular reserve which he has for exercise.

As regards the expression of our specific haemoglobin flow in units, it can be obtained in the same units which we have already used for other things, for the haemoglobin flow in terms of oxygen is the blood flow multiplied by the oxygen capacity:

$$\text{S.H.F.} = \frac{\text{B.F.} \times \text{O.C.}}{\text{D.C.}}.$$

But

$$\text{B.F.} = \frac{\text{M.R.}}{\text{A.O.T.}},$$

$$\text{O.C.} = \frac{\text{A.O.T.}}{\text{R.O.T.}},$$

$$\text{D.C.} = \frac{\text{M.R.}}{\text{D.P.H.}};$$

$$\text{S.H.F.} = \frac{\text{M.R.}}{\text{A.O.T.}} \times \frac{\text{A.O.T.}}{\text{R.O.T.}} \times \frac{\text{D.P.H.}}{\text{M.R.}} = \frac{\text{D.P.H.}}{\text{R.O.T.}}.$$

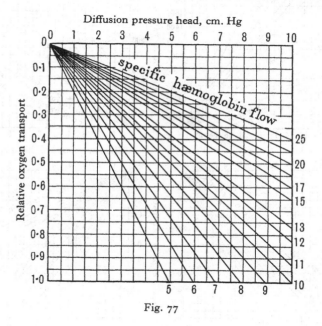

Fig. 77

Now I must make a confession. Although apparently it is a sheer matter of arithmetic—as is the fact that s.h.f. can be expressed in terms of D.P.H. and R.O.T.—the mental picture is always to me more obscure than the actual one (Fig. 77). But apparent or obscure, the convenience of this relation is quite extraordinary, for it enables us to build up a picture of the man in terms of the four relations which we have considered. Placed singly in juxtaposition they appear as in Fig. 78. It will be seen that they piece together into a single figure which we may now proceed to turn to account (Fig. 79), as was first shown by Murray and Morgan (1925).

Let us commence, therefore, by depicting a normal resting man

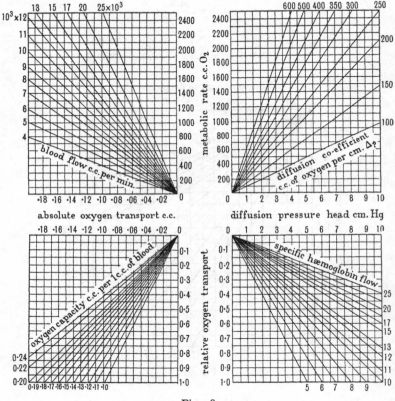

Fig. 78

on the above background. Let me take my old colleague, A. V. Bock, the figures given by him having been obtained with exceptional care and being very complete.

I said the *resting* man. That gives me my starting-point, for the metabolic rate is the measure of the activity of the individual and is the obvious factor which, while preserving his normality can be altered voluntarily. The metabolic rate is 253 c.c. per min. Bock's blood flow was also measured and was 4900 c.c. per min. From the metabolic rate and the blood flow the absolute oxygen

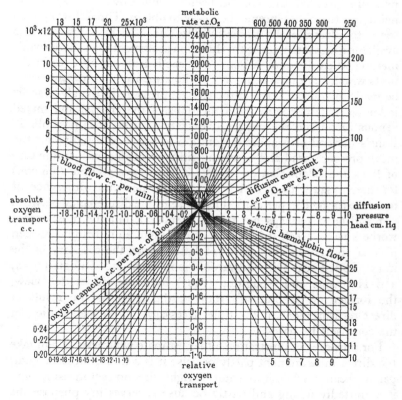

Fig. 79

transport is directly deducible and is 0·051 c.c. The third measurement, which is easily made, is the total oxygen capacity. That proved to be 21·7 volumes of oxygen per 100 volumes of blood. Hence we obtain a fourth point on the figure which gives the lower left-hand corner of the quadrangle, and also the point on the ordinate corresponding to the percentage oxygen transport, which is 0·23, and so we are half way round the quadrangle and in the fortunate position that one more measurement will give us the rest —be that measurement either the diffusion coefficient, the diffusion pressure head, or the specific haemoglobin flow. Unfortunately neither of these measurements is easy to obtain. The following data serve to illustrate the difficulty. In the case of Bock there is available a measurement of the diffusion coefficient made by Krogh's method when we were in Peru. The coefficient was 27·1 (or in the units which I am now using 271). The measurement may however be made in another way. If the oxygen pressure in the alveolar air is known together with the saturations of the arterial and mixed venous blood and the oxygen dissociation curve, it is possible by Bohr's method of graphic integration to arrive at a figure for the mean pressure head and so for the diffusion coefficient. In the case of Bock the calculation on one set of data gave 38 mm. (3·8 cm.) for the pressure head and 6·7 c.c. of O_2 per mm. difference of pressure (or 67 c.c. per cm.). The cause of the discrepancy between 271 and 67 is perhaps not fully understood, but the most recent workers would seek it in a lack of complete equilibrium between plasma and corpuscle before the latter leaves the capillary of the lung. Both Bock, Dill, Edwards, Henderson and Talbott (1929) and Hartridge and Roughton (1927) have arrived at this view, the former on theoretical grounds, the latter as the result of direct measurement of the time necessary for the penetration of the corpuscle by gases.

For purposes of illustration either estimate will serve. I will take the direct measurement partly because it is a direct measurement, partly because, while it may not be right, the indirect measurement is admittedly wrong and partly because it serves my purpose the less well of the two. Provisionally then we place the diffusion coefficient at 271 c.c. of oxygen per cm. difference of pressure. The fourth side of the rectangle is now complete.

It will be convenient to consider the factors altered during exercise by the method described by Murray and Morgan.

The fact has already been stated that, of the various factors plotted on Murray and Morgan's chart, the metabolic rate is the one which can be altered at will. Any man up to the prime of physical life and of good physique can increase his oxygen consumption from under 300 to over 3000 c.c. per min., and doubtless many men can raise it to a much higher level. Let us consider the question in the light of the statement at the head of the chapter "Every adaptation is an integration". Let us commence by considering the state of affairs if, on exercise, nothing at all changed except the increase of oxygen used.

We may take the figures given by Bock for exercise corresponding to a metabolic rate of 2400 c.c. per min. Let us now go round the diagram (Fig. 79) counter-clockwise: the first thing to which we come is the blood flow. Bock's resting blood flow is in the region of 5 litres a min., but 5 litres of blood cannot take up 2400 c.c. of oxygen. To do so the absolute oxygen transport would have to be 0·48 c.c.—quite off the picture and necessarily so because the oxygen transport cannot exceed the oxygen capacity, and the oxygen capacity is only 0·21 c.c. The blood flow therefore must quicken. But to what extent? Let us see what will happen if it quickens just so much as to keep the next factor to which we come —the absolute oxygen transport—constant at a value of 0·05 c.c. per c.c. of blood. The blood flow per minute must needs be 48 litres to preserve the absolute oxygen transport constant. With all due respect to one of my best friends, let me say that his heart is incapable of that feat. Forty-eight litres is around the capacity of twenty Winchester quart bottles. Try to fill a Winchester quart bottle from a tap in 3 seconds and you will exonerate me from trying to belittle the physique of my friend. Therefore the absolute oxygen transport must come to some new position greater than at rest, yet within the limits imposed by his oxygen capacity. When I say the limits imposed by the oxygen capacity I mean, in the first instance, that the relative oxygen transport must not rise above 1; and that brings us into the second quadrant of our picture. In point of fact the relative oxygen transport cannot afford to be increased to nearly so great a degree, because the ultimate object of oxygen transport is the feeding of the tissues. The escape of oxygen from the blood being a matter of diffusion demands a considerable partial pressure in the capillary which, in its turn, demands incomplete reduction. It would seem that proper nutrition of the tissues is not consistent with an oxygen saturation of less than 40 per cent. in the general venous blood, which would mean a relative oxygen transport of about 55 per cent. or in our units 0·55. Now just as a more rapid blood flow will keep down the absolute oxygen transport when the metabolic rate rises, so an increase in the total oxygen capacity will keep down the relative oxygen transport when the absolute oxygen transport rises. It is not surprising, therefore, to discover that on exercise there is a rise in the oxygen capacity of the blood, *i.e.* in the total oxygen in each cubic centimetre of the circulating fluid.

Having completed the left half of the rectangle, let us now turn to the right half and treat it in the same way. What will occur if the diffusion coefficient remains as at rest, in the vicinity of 271 c.c. per cm.? The diffusion pressure head would become 2400/271, roughly 8·8 cm. or 88 mm. Clearly it is as impossible to have a pressure head of 88 mm. For the alveolar oxygen pressure is at most 120 mm. during exercise, and that in the venous blood about 30 mm.; so that 88 mm. would represent approximately the difference not between the pressures in the alveolar air and the average of the capillary, but between the alveolar air and the vein. Therefore, the diffusion coefficient—if indeed it were 271 at rest—must increase materially on exercise. But to what extent? Can it increase so much as to leave the diffusion pressure head unchanged at say about 9 mm. (0·9 cm.)? That would mean a diffusion co-efficient of about 2670 c.c. of oxygen per cm. of pressure head. This tenfold increase would appear most improbable. On the assumptions that the arterial blood is 95 per cent. saturated, the venous blood 40 per cent. saturated, and the alveolar oxygen pressure 120 mm., the mean pressure head is 70 mm. (7 cm.) and the diffusion coefficient 35, or with our units 350. I do not argue for the correctness of this figure, but it will serve. The immediate point is that the diffusion coefficient by calculation takes up an intermediate position around 350.

We have now been round the whole quadrangle, and have seen that alteration in the metabolic rate causes an alteration in each of the seven other quantities which compose the figure. With increase in the metabolic rate each other factor increases to some extent; otherwise, in the words of A. V. Hill, the system would jam. This is what I mean by an adaptation being an integration.

Nor is this all. Murray and Morgan pointed out that from the viewpoint of "cause and effect" the seven quantities which accommodate themselves to the increased metabolic rate were not all on an equal level, and I have expressed the same fact by speaking of one at least as a "matter of arithmetic". Going round the quadrangle it is evident that, apart from the metabolic rate, there are three factors which are the direct result of physiological changes invoked in the activity of the organism. These are formed at three corners. They are the rate of blood flow, the oxygen capacity of the blood, and the diffusion coefficient. So far, therefore, as bodily function goes, the whole alteration in the quadrangle consequent on the altered metabolic rate is an integration of the changes in these three active variables. But is either alteration of the minute volume, or the oxygen capacity, or the diffusion coefficient, a simple process? Is not each itself an integration? It is surprising how little we know about any of the above, but, briefly to review what is known, consider first the oxygen capacity of the blood. The alteration which has been observed in man is up to about 10 per cent. of the total oxygen capacity, not more (Table XVIII). It was first observed by Himwich and Barr (1923), and has been verified by Bock and his colleagues (1928) and by Harrison, Robinson and Syllaba (1929).

Table XVIII. *Oxygen capacity of* 100 *c.c. blood*
(*Himwich and Barr*, 1923)

	Before exercise c.c.	During exercise c.c.	After exercise c.c.
D. P. B.	20·5	—	21·9
H. E. H.	19·1	20·0	20·2
D. P. B.	21·3	—	23·8
D. P. B.	22·1	—	23·7
H. E. H.	21·1	—	21·6

In some animals the change appears to be much greater, possibly because the animals were subjected to a more considerable degree of exercise, but I think more probably because such creatures as the horse have been developed along the lines of great capacity for physical exertion. There are two principal known reactions, either of which would tend to raise the red corpuscle count: the one is the contraction of the spleen; the other is the imbibition of water by the tissues. The former is considered more particularly in the chapter on material stores. Here it need only be said that a dog of 18 kg. appears capable of storing in the spleen about 1,500,000,000,000 red corpuscles which can be expelled into the general circulation with a less proportion of plasma than is present in ordinary blood. But it is clear that the disparity between the corpuscles and the plasma in the blood from the spleen pulp is not so great as to account for the great concentration of corpuscles in the blood of the general circulation. The evidence that the spleen in man contracts on exercise is based on X-ray photographs. Opinions differ as to whether the observed increase in the blood corpuscle count does or does not take place in splenectomised men. The second factor which makes for concentration of the blood is the rise in osmotic pressure of the contracting muscles. One of the oldest experiments in the physiology of muscle is that which shows that a muscle if made to contract in oil does not change in volume. It is difficult to see how it could do so, for there can be no accession of material into the substance of the muscle under such conditions, but the case is very different when muscle is made to contract in a medium from which it can imbibe water. There is of course a large literature on this imbibition of water by excised muscles during contraction in a medium of saline, but even more to the point were the observations of Back, Cogan and Towers (1915), and of Barcroft and Kato (1915) on the imbibition of water by muscles made to contract in the body and fed meanwhile with their normal circulation. According to an old observation of Ranke (1880), quoted by Gamgee: "The amount of solid matter in muscles undergoes diminution when muscles are tetanized, so that there appears to be a relative increase in water".

Table XIX is among those given by Back, Cogan and Towers. In the three controls in which neither muscle was stimulated

there was at most a difference in weight of 1·3 per cent. between
the left and right muscles, whilst in the experiments in which one
side was stimulated the muscle on that side was up to 7·6 per cent.
heavier.

Table XIX. *Weight of frog's gastrocnemius*
(Back, Cogan and Towers)

Left side mg.	Right side mg.	Difference mg.	Difference %
340·5	340·5	0	0
146·0	147·5	1·5	1·0
116·0	118·0	1·6	1·3

Unstimulated mg.	Stimulated mg.	Gain on stimulated side mg.	Gain on stimulated side %
462·0	487·0	25·0	5·1
345·0	353·0	8·0	2·3
219·5	232·7	13·2	7·6
310·0	329·8	19·8	6·1
427·0	430·0	3·0	0·7

That the difference in weight is due to passage of water into the
muscle either as the result of osmotic phenomena or of alteration
of the permeability in the capillary wall is shown by the specific
gravity determinations made by these authors (Fig. 80 and
Table XX).

Table XX

Average weight of muscles on two sides mg.	Excess weight of stimulated muscle mg.	Sp. gr. of stimulated muscle	Sp. gr. of unstimulated muscle
192	37·0	1·061	1·073
257	25·0	1·064	1·075
235	7·6	1·075	1·078
237	3·2	1·060	1·066
323	28·4	1·070	1·069
437	7·2	1·069	1·075
436	5·8	1·069	1·072
281	6·1	1·068	1·071
277	27·6	1·073	1·087

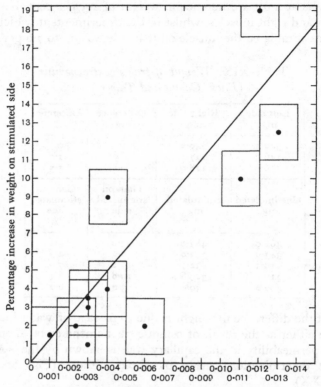

Decrease in specific gravity of stimulated muscle
Fig. 80

As regards oxygen capacity, Table XXI gives the percentage increase in oxygen capacity of blood traversing the resting and stimulated muscles respectively.

Table XXI. *Percentage increase in oxygen capacity*

Exp.	Resting	Stimulated
1	1·5	1·5
2	1·9	3·4
3	0·6	4·6
4	1·0	2·8

Not all the water leaving the blood is retained by the muscle, some leaves as lymph. Nevertheless, to take a single case, the weights and specific gravities of the gastrocnemii of a dog after severe stimulation on one side are given in Table XXII. Considering the fallibility of the methods employed (principally the possibility of the muscles on the two sides being occasionally of markedly different weights to start with) it seems a fair assumption that the increase in weight of the muscle on the stimulated side is due to transference from the blood. The blood, of course, might be expected to make up the loss to some extent by itself depriving non-stimulated tissues of water. Hence a marked effect on the water content of the blood could only be expected when the active tissues form a large proportion of the soft parts of the body.

Table XXII

	Stimulated side	Unstimulated side
Weight (gm.)	38·2	31·6
Specific gravity	1062	1073
Calculated weight of stimulated muscle (gm.) ... (From alteration sp. gr.)	36·9	—

The observations of Barcroft and Kato on mammalian muscle show (1) that the blood traversing the muscle loses water on a much greater scale when the muscles are stimulated than when they are at rest, (2) that the muscle on the stimulated side is markedly heavier, and (3) that its specific gravity is smaller.

The relative importance of the spleen in raising the total oxygen capacity during exercise seems to depend upon the animal; in man it is uncertain (Dill, Talbott and Edwards, 1930; Benhamou, Jude, Marchioni and Nouchy, 1929); in the horse, dog and cat it is considerable. The question has been studied by Scheunert and Krzywanek (1926). These authors in the research in question worked upon the horse. They judged of the increase in total oxygen capacity by the erythrocytic count and by haematocrit readings, and of the water lost to the tissues by estimation of the serum proteins. By way of preface two points may be noted: (1) that the quantity of proteins in the plasma from the spleen pulp is the same as that in the circulating plasma, and (2) that the salts in the plasma are not altered in concentration by the taking of exercise. The implication contained in these two points is from the first that increase in the proteins of the plasma is not due to extra protein put in from the spleen, but to water which has left the blood; and from the second that such water carried the salts with it and is therefore an osmotic effect by which water and salts are carried through the capillary wall. Having said so much, we may pass to Scheunert and Krzywanek's results, of which Table XXIII is an example.

Table XXIII

Date (1926)	Condition when blood was drawn	Haematocrit	Leucocyte count	Serum proteins %	Crystalloids in serum %
Feb. 2nd	Rest	24	7440	6·16	1·5
	5 min. trot	35·5	8960	6·77	1·44
Feb. 5th	Rest	25	8180	6·82	1·13
	5 min. trot	37	8760	7·28	1·13
Feb. 17th	Rest	29	—	6·21	1·63
	5 min. trot	39·5	—	6·62	1·63
Average increase on exercise expressed as a percentage of resting value		43	13	8	−1·5

Taking the experiment of February 2nd, 1926, Scheunert and Krzywanek assume the horse in question to have a blood volume of 30 litres. It would therefore have a plasma volume of 22·8 litres and a corpuscle volume of 7·2 litres. Judging from the difference in concentration of the plasma proteins the plasma volume would be reduced on exercise to $\frac{22 \cdot 8 \times 6 \cdot 16}{6 \cdot 77} = 20 \cdot 7$ litres. The corpuscle value would then become 7·2 litres in 27·9 or approximately 26 per cent. instead of 24. Yet it was actually 35 per cent. What can the spleen contribute towards the balance? The blood in the splenic vein contained 61 per cent. of corpuscles. Of such blood

10 litres, if added to 28 litres which already contained 26 per cent. of corpuscles, would make the whole up to 35 per cent. The spleen after exercise would then have been only one-fifth of its original size.

What other possible mechanisms are there by which either water may be withdrawn from the blood or corpuscles may be added to it within 5 minutes? Considerable stress has been laid by Lamson (1915, etc.) on the contribution of the liver in this respect. His view is that the concentration of the blood which occurs on the administration of adrenaline is due in part to the abstraction of water which goes to form lymph, and in part to the expulsion of blood from the liver capillaries. The difficulties about this theory seem to me, firstly, that the amount of lymph postulated appears to be much greater than the liver capillaries can hold; any lymph which goes back into the blood of course cancels just to the same extent further withdrawal in the liver: secondly, that other observers such as Binet maintain that in the absence of the spleen adrenaline does not produce concentration of the blood. Thus we have two extreme views of adrenaline concentration: Lamson's view that the liver is the sole cause of blood concentration, and the view of Binet and his collaborators (1927) that the spleen completely accounts for the polycythaemia. The latter view seems to have been confirmed by Izquierdo and Cannon (1928) so far as the cat is concerned. In man there is a difference of opinion. Dill, Talbott and Edwards state that there is no contraction, while Benhamou, Jude, Marchioni and Nouchy quote X-ray observations to establish the contraction of the spleen in man as the result of exercise. Granting the correctness of their observations, the issue would be as to whether the material expelled was richer than ordinary blood in erythrocytes or was not. In the former case the rise in the blood count should be greater in normal than in splenectomised persons; in the latter case it should not be so.

To turn now to the blood flow, is the augmentation of the blood flow the result of some simple action of the body or is it too an integration? It is surprising how little is known about a matter so fundamental. It is not possible to give any complete account of the mechanism by which the blood flow can be raised from 5 to 25 litres per min. The following factors may be considered: the heart rate; the injection of fluid into the great veins; the effect of respiration; a redistribution of blood.

The heart rate. Some 20 years ago, had the question "What causes increased blood flow?" been asked, it would probably have received the answer "The increased pulse rate". That answer was very natural; the heart being regarded as a pump, what could seem more likely than that the more frequent the strokes of the pump, the greater the quantity of fluid would be passed through it. The information given by the heart-lung preparation however did not agree with that conception. It appeared that for all moderate blood flows the amount of blood which traversed the heart depended solely on the pressure at which it was fed into the venous side and not at all on the rate of the heart beat. The recent work of Henry Barcroft (1932), working in Anrep's school, seems to indicate the possibility of a shift of opinion in the direction of the older school of thought. What is true of the heart-lung preparation is not necessarily true of the closed circuit of the body. In the intact animal it seems that the injection of adrenaline either into a vein or directly into one of the chambers of the heart sometimes causes an increase in the blood flow coincident with the quickening of the pulse. It must not be supposed, however, that the increased pulse rate is the sole cause of the increased blood flow, for if the heart be quickened by electrical stimulation a comparable increase in blood flow does not occur. Indeed even this one factor appears to be an integration of the combined effects of adrenaline on the rate of the heart and on the condition of its muscle.

Injection of fluid into the great veins. The cause of increased blood flow has been studied by Daly (1925) in a form of heart-lung preparation which, unlike the Starling preparation, was a closed circuit. Daly obtained the very striking result that injection of quite small quantities of blood produced very considerable increments of the flow round the body.

The results of one experiment are shown in Fig. 81. The experiment ran the following course: starting with a blood flow of 0·53 litre per min., the injection of 20 c.c. of blood was made, followed by six injections each of 10 c.c. and a final injection of 20 c.c., making 100 c.c. in all, by which time the blood flow had reached 1·39 litres per min. 100 c.c. of blood was then withdrawn in five portions, each of 20 c.c., and the blood flow returned to 0·71 litre.

With regard to the question of whether the rate at which the injection was made had much influence on the increase of the circulation rate in the dog, 35 c.c. of saline when injected in 30 sec. produced an immediate rise in the circulation rate from 150 to 330 c.c. per min. Injected more rapidly the same quantity of fluid

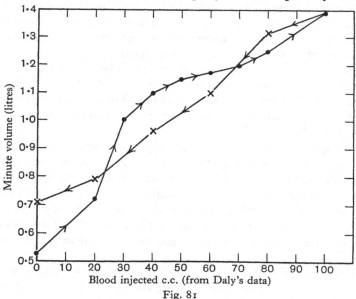

Fig. 81

produced a greater immediate rise, namely, from 195 to 455 c.c. per min. In either case the great rise was evanescent; the circulation steadied down to a condition in which the circulation rate was 41 per cent. greater than it was to start with. It did not appear therefore that, at the rates of injection employed, the actual rate made much difference to the final results.

The fact which stands out from Daly's experiments is that injection of a quantity of blood so small as 100 c.c. may increase the minute volume in his preparation two and a half times.

As to where 100 c.c. of blood can come from in the actual animal, the best attested place outside the circulation is the spleen pulp. There is little doubt that the spleen of a dog of 10 kg. could put a considerable fraction of 100 c.c. into the circulation at the commencement of exercise or even before the actual onset of muscular work. Here, however, I must issue a caution lest too much be expected of this organ whose function a while ago seemed so obscure. I mention the spleen here simply because I know that it evacuates itself partially at least at the commencement of exercise, having seen it do so many times. Blood probably comes from other organs in ways only now becoming understood (see Chapter VI).

The effect of respiration on the blood flow. It is a commonplace that in an animal whose heart has ceased to beat, a circulation of meagre character can be maintained by the application of artificial respiration. If, then, in the normal animal there is a transition from quiet to deep respiration, what effect, if any, will the change produce upon the blood flow?

This question also has been the subject of research by Daly on his closed heart-lung preparation. He found that by placing the heart and lungs in a box to which suction could be applied, in imitation of deep respiration, the circulation rate could be increased 150 per cent. At each inspiration the diastolic pressure in the chambers of the heart was reduced, blood was therefore sucked into the heart along the veins and so the cardiac output was increased.

A redistribution of blood. Certain possibilities by which a redistribution of the blood can increase the rate of flow through the heart have been revealed and investigated in the Cambridge laboratory by Henry Barcroft (1931 *a*, *b*). The work has been carried out on dogs, and has been made possible by the invention of an automatic stromuhr which can deal with a blood flow of more than a litre per minute. The output can be measured with accuracy from moment to moment (each 25 c.c. being recorded) and with a minimal resistance to the circulation.

As so frequently happens when some new factor becomes readily measurable, paradoxical results commenced to appear. Perhaps the most important was as follows: the stromuhr being introduced at F (Fig. 82), when the circulation was clamped at B, the flow through the stromuhr frequently quickened. This phenomenon apparently depends upon the circulatory system possessing:

(*a*) A heart which has the property of putting out all the blood it receives irrespective of the pressure against which the blood is expelled.

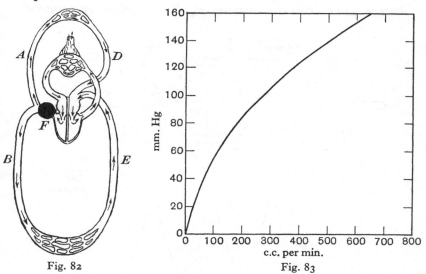

Fig. 82   Fig. 83

Fig. 82. Diagram of animal's vascular system: A, brachiocephalic artery; B, thoracic aorta; D, superior vena cava; E, inferior vena cava; F represents the position of the stromuhr; F represents the position of the aortic injections of adrenaline; B represents the point at which the screw clamp was applied; E represents the position of the injections of adrenaline into the femoral vein.

(*b*) A capillary system which can hold a considerable volume of blood, and is imperfectly elastic, *i.e.* equal increments of pressure in the system are produced by successively decreasing quantities of blood put into it.

That a healthy heart possesses property (*a*) was of course shown by Patterson and Starling (1914) on the heart-lung preparation; property (*b*) remains for consideration. Fig. 83 shows the relation of the rate of blood flow through the brachiocephalic artery to the pressure in the artery.

To imitate the capillary circulation a model which has the properties stated may be introduced into the heart-lung preparation. It consists of three vertical tubes, each conical in shape, tapering towards the top.

The blood from the heart passes through a series of perforated rubber diaphragms, R_1, R_2, R_3, which represent the arterial resistance, and another similar series, R_4, R_5, R_6, which represent the

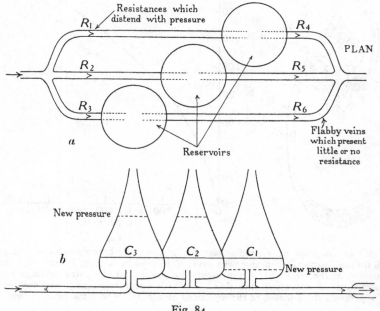

Fig. 84

resistance of the capillaries. Between the two series of diaphragms are vertical tubes, C_1, C_2, C_3, conical in shape, tapering towards the top (Fig. 84). At the commencement of the experiment blood stands in the tubes C_1, C_2, C_3 at a level about that of the normal capillary pressure (25 cm.). If now a clamp is placed at R_1 the blood runs out of C_1, the pressure rises in C_2 and C_3 to a degree which is disproportionate to the volume of blood in these vessels and the volume of blood passing into and out of the heart increases.

Let us then assume that we have discovered four ways in which the minute volume may be increased:

(1) By the effect of sympathetic stimulation and vagus inhibition, say twofold.

(2) By the effect of injection of blood from some backwater or backwaters into the veins, say two and a half-fold.

(3) By effect of deeper respirations, say twofold.

(4) By a redistribution of blood.

What is the combined effect of these influences? To put the matter bluntly, are we justified in multiplying these factors together? That indeed would be a daring form of integration. It would give us more than a tenfold quickening of the blood flow. As far as the third point cited above (the effect of increased depth of respiration) is concerned, it looks like an arrangement for amplifying the flow, whatever it may happen to be, in a definite ratio rather than like a scheme for adding a definite quantity to the pre-existing minute volume.

But whether the increments are made the subject of addition or multiplication, they form a very striking example of team work on the part of the body. Apart from the brain, which is at the back of all such changes, the heart itself is involved in one factor, the spleen and possibly other viscera in a second, and the muscles of respiration in a third. There we may leave the augmentation of the blood flow.

The diffusion coefficient forms the third corner of Murray and Morgan's rectangle. About it even less is known than about the two which we have discussed. The original method of measuring the diffusion coefficient is that of estimating the amount of carbon monoxide which passes into the blood in a given time, which time must be less than that taken for the blood to pass round the body. When the pressure of CO in the blood is nil throughout the whole operation, its pressure in the alveolar air is determined, as also is the quantity of the gas which disappears from the lung. The diffusion coefficient for CO being known, that for oxygen is inferred from the relative known rates of diffusion of the two gases.

No diffusion coefficient for Bock during exercise has been obtained in this way. It was calculated from the metabolic rate, the oxygen pressure in the alveolar air and the estimated average oxygen pressure in the lung capillary.

It is a happy coincidence that Bock is perhaps that one person, data for whose coefficient at rest are available by both methods. It is of interest to compare them. When in Peru in 1922 Bock's diffusion coefficient was measured on two occasions by Harrop, the method being that of CO absorption. It gave

At Lima   27·1
At Cerro de Pasco 38·8

These figures were, let me repeat, for Bock *at rest*. They differ but little from one another, the former is that given on the chart. At rest Bock's diffusion coefficient, as calculated from the combined measurements of the oxygen pressures in his alveolar air and arterial blood and his metabolic rate, is of the order 6, whilst at full exercise it is 41. So great a difference as between rest and exercise is quite at variance with Krogh's work. Krogh found the difference relatively trifling. Presumably the slightly higher diffusion co-efficient on exercise was due to the capillary walls becoming on the average thinner, with the greater average distension of the lung.

It is possible however that neither diffusion coefficient is correct, for doubt has been shed by the most recent workers (Hartridge and Roughton (1927), Roughton (1925), and Bock, Dill, Edwards, Henderson and Talbott (1929)) on the fundamental assumption which rules both measurements, namely, that the whole system is in equilibrium when the blood leaves the lung. The latter workers suggest that oxygen is absorbed from the plasma after the blood has ceased to have access to the alveolar air, whilst Roughton considers that even the higher estimates put forward by Krogh are too low on some similar grounds which apply more particularly to carbon monoxide. If such errors exist during rest, they probably loom larger during exercise, but whatever latitude be allowed, it seems impossible to escape from the conception illustrated in Fig. 79 that the adaptation to exercise is an integration of a large number of factors no one of which could alter sufficiently to be completely effective.

REFERENCES

BACK, M., COGAN, K. M. and TOWERS, A. E. (1915). *Proc. Roy. Soc.* B, **88**, 244.

BARCROFT, H. (1931 *a*). *J. Physiol.* **71**, 280.

—— (1931 *b*). *Ibid.* **72**, 186.

—— (1932). *Ibid.* **76**, 339.

BARCROFT, J. and KATO, T. (1915). *Phil. Trans. Roy. Soc.* B, **207**, 149.

BENHAMOU, E., JUDE, MARCHIONI, and NOUCHY. (1929). *C.R. Soc. Biol.* **100**, 456.

BINET, L., CARDOT, H. and FOURNIER, B. (1927). *Ibid.* **96**, 521.

BOCK, A. V., DILL, D. B., EDWARDS, H. T., HENDERSON, L. J. and TALBOTT, J. H. (1929). *J. Physiol.* **68**, 277.

BOCK, A. V., VANCAULAERT, C., DILL, D. B., FÖLLING, A. and HURXTHAL, L. M. (1928). *Ibid.* **66**, 136.

DALY, I. DE B. (1925). *Ibid.* **60**, 103.

DILL, D. B., TALBOTT, J. H. and EDWARDS, H. T. (1930). *Ibid.* **69**, 267.

HALDANE, J. S. (1901). *Ibid.* **26**, 497.

HARRISON, T. R., ROBINSON, C. S. and SYLLABA, G. (1929). *Ibid.* **67**, 62.

HARTRIDGE, H. and ROUGHTON, F. J. W. (1927). *Ibid.* **62**, 232.

HIMWICH, H. E. and BARR, D. P. (1923). *J. Biol. Chem.* **57**, 363.

IZQUIERDO, J. J. and CANNON, W. B. (1928). *Amer. J. Physiol.* **84**, 545.

JENKINS, C. E. and DON, C. S. D. (1933). *J. Hyg.* **33**, 36.

KROGH, M. (1915). *J. Physiol.* **49**, 271.

LAMSON, P. D. (1915). *J. Pharm. Exp. Therap.* **7**, 169.

—— (1916). *Ibid.* **9**, 129.

—— (1920). *Ibid.* **16**, 125.

LAMSON, P. D. and KEITH, N. M. (1916). *Ibid.* **8**, 247.

MURRAY, C. E. and MORGAN, J. (1925). *J. Biol. Chem.* **65**, 419.

PATTERSON, S. W. and STARLING, E. H. (1914). *J. Physiol.* **48**, 357.

RANKE, J. Quoted in *A Text-book of Physiological Chemistry of the Animal Body*, A. Gamgee, **1**, 364. London, 1880.

ROUGHTON, F. J. W. (1925). Dissertation for Ph.D. Degree, Cambridge University.

SCHEUNERT, A. and KRZYWANEK, F. W. (1926). *Pflügers Arch.* **213**, 198.

CHAPTER IX

EVERY ADAPTATION IS AN INTEGRATION
(*continued*)

ADAPTATION TO ANOXIA*

The condition of insufficient supply of oxygen to the tissues, for that is what anoxia has come to mean, may be classified under three main headings (Barcroft, 1920–1).

The anoxic type in which the arterial blood is insufficiently saturated.

The anaemic type in which the oxygen capacity is abnormally low.

The ischaemic type in which the quantity of blood circulating is less than normal.

There are other more obscure conditions the relationship of which to anoxia has been considered by Peters and van Slyke (1931), *e.g.* cyanide poisoning. In cyanide poisoning the abnormality, however, lies not in any fault of transport but of the ability of the tissues to use the oxygen supplied to them.

Taking the three obvious types, each is capable of representation on Murray and Morgan's diagram and therefore each may be considered with regard to it.

The anoxic type, once it reaches the point of anoxia, finds its expression primarily in a reduction in the diffusion pressure head, the anaemic type primarily as a reduction in the oxygen capacity, and the ischaemic type as a reduction in the blood flow. It will be well to consider each case with regard to a standard metabolic rate, though there is no doubt that each deficiency ultimately expresses itself as a reduction of the maximum metabolic rate of which the organism is capable.

For a given metabolic rate, however, each disorder may be con-

---

* In agreement with Peters and van Slyke, I am using the word "anoxia" in place of "anoxaemia". "Anoxia" signifies any condition which retards the oxidation processes in the tissues.

sidered as a condition in which the rectangle is driven inwards at some one point. The questions which we have to consider are:

(1) Does adaptation exist?

(2) If so, is it an integration of several factors?

As the anoxic type of anoxia is that of which we know most, it will be the first to be considered.

Adaptation to anoxic anoxia has been the subject of much laborious work from many points of view, primarily exposure to abnormally low oxygen pressures such as are experienced at high altitudes. It occurs, however, in many diseased conditions, such for instance as inflammation of the lungs, patent foramen ovale, emphysema, etc. The difficulty with regard to such conditions is to separate the effects of anoxia, pure and simple, from the concurrent and, to the organism, probably much more important effects following from collateral disturbances such as toxic poisoning.

ADAPTATION TO ALTITUDE

It would be of great interest to express the factors in the adaptation to altitude and other conditions in precisely the same way as has been done for exercise. The difficulty in doing so lies in the lack of precise data for the purpose. But though there are wide gaps, which must simply be admitted as such, the picture is worth drawing if only because the imperfections in the quadrangle make quite clear the fact that the adaptation to altitude is an integration. A future generation can carry out the integration correctly; my present purpose goes no further than to show that the process of adaptation is integrative.

The essence of the problem lies in the following facts:

(1) The pressure of oxygen in the pulmonary alveoli is normally about 100 mm. Hg and is therefore about 60 mm. below that in the atmosphere. Under these circumstances, the saturation of the arterial blood with oxygen is taken as 95 per cent.* and that of

---

* The more recent determinations indicate that the saturation of the blood at rest is usually definitely below 97 per cent., the figure obtained by equilibrating blood with alveolar air. The latter figure is more nearly obtained during hyperventilation.

the venous blood 60 per cent. (Fig. 85, X and Y respectively), and
the unloading is 35 per cent.

(2) At such a place as Cerro de Pasco, 15,000 feet above sea-
level, the barometer stands at 450 mm., of which 95 mm. is oxygen.
If no form of adaptation occurred, the alveolar pressure would be
95 − 60 = 35 mm.

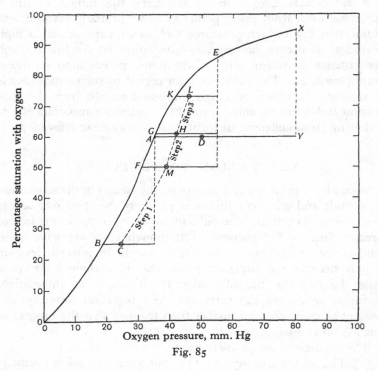

Fig. 85

(3) Under these circumstances the arterial blood would be not
more than 60 per cent. saturated (Fig. 85, A) (assumption 1 made
here and from now onwards, that with the greater expansion of the
chest the arterial blood is approximately in equilibrium with the
alveolar air) and the venous blood if 35 per cent. less saturated
than the arterial would be only 25 per cent. saturated and the
oxygen pressure would be 18 mm. (Fig. 85, B).

(4) The mean pressure of oxygen in the capillary. (Assumption 2. The mean capillary pressure is reckoned as follows: the difference in the pressure of oxygen in the artery and vein is divided by three and the figure so obtained is added to the pressure in the vein. Thus in the present case the pressure in the artery is 35, that in the vein 18—difference 17 mm.—the capillary oxygen pressure is therefore taken as 24 mm. (Fig. 85, C). This treatment is purely empirical and can only be used where the dissociation curve is steep and the oxygen pressure in the tissues considerably below that in the veins, nor is any great accuracy claimed for it, but it will suffice for purposes of the present argument.) 24 mm. is now only about half the normal value (Fig. 85, D) and is too low for the efficient working of the organism. How is it to be raised? The answer so far as it can be given is as follows.

Step 1. The first step is, if possible, to raise the pressure of oxygen in the alveolar air. This can be, and is, accomplished by an increase in the total ventilation. Granting that a certain quantity of oxygen is taken out of the air, the more air which ventilates the lungs the higher will be the oxygen pressure in the alveoli.

By that process at Cerro de Pasco the oxygen pressure in the alveolar air of our party was raised to an average of about 55 mm. This alteration meant that the limiting saturation of the arterial blood would rise to about 85 per cent. (Fig. 85, E) and in the absence of any other change that of the venous blood to 50 per cent. saturation (F). The resulting average oxygen pressure in the capillaries would be 39 mm. (M). Step 1 then means a rise from 24 to 39 mm.

Step 2. The second factor in adaptation to high altitudes which I shall consider, though not the second to supervene in order of time, is increase in the haemoglobin value of the blood. This is a factor which differs considerably in different persons. Suppose however, as in Haldane's case, the haemoglobin increases from 100 to 150 on the haemoglobinometric scale. The repercussion on the unloading of oxygen will be as follows: the same absolute quantity of oxygen will be unloaded, but that quantity expressed as a percentage of the oxygen capacity will be very different; whereas the blood unloaded 35 per cent. of its oxygen in going round the body, it will now unload only $35 \times \dfrac{100}{150} = 24$ per cent. Then if the arterial blood was 85 per cent. saturated the venous blood would be $(85 - 24) = 61$ per cent. saturated (G). The pressures corresponding to these saturations are 55 and 36 mm. respectively and the capillary oxygen pressure would on our reckoning be 42 mm. (H).

Step 3. The third step is an acceleration of the blood flow so that the minute volume of blood which circulates is increased. Let us assume that the minute volume is doubled. If the absolute quantity of oxygen unloaded from the blood is unchanged, the percentage unloading will be halved from its value at the end of Step 2, and now therefore becomes 12 instead of 24 per cent., and with the saturation of the arterial blood 85 per cent. (once more assuming that the arterial blood is approximately in equilibrium with the alveolar air) that of the venous blood will be $85 - 12 = 73$ per cent. (K), and the corresponding capillary oxygen pressure will be 46 mm. (L). Summarising then we get the following rough figures:

Table XXIV

	Mean capillary oxygen pressure (mm.)
Normal	50 (D)
15,000 ft., no adaptation	24 (C)
Step 1	38 (M)
Step 1 + Step 2 ...	42 (H)
Step 1 + Step 2 + Step 3	46 (L)

and so as the result of these three steps we have arrived at nearly the normal figure, but with an organism which has drawn on its reserves, to the extent of breathing more deeply, possessing blood of higher viscosity and having a twofold minute volume.

Actually at the altitude under consideration the change is less serious than represented because, according to Grollman (1930), Step 3 (the increased minute volume) is at this altitude somewhat of an emergency method, as it appears at once after the ascent and passes off in time—presumably as the increase of haemoglobin becomes more marked (Fig. 86).

At a higher altitude doubtless both conditions occur together. But at higher altitudes there are further factors in adaptation. Below 15,000 ft. there is no difficulty in saturating the blood with oxygen to a high degree in the lungs—the difficulty has been to do so and at the same time maintain a venous oxygen pressure high enough for the purposes of the body, notably the brain—but at higher altitudes a point must be reached at which even with considerable hypernoea the arterial blood would fail sufficiently to saturate itself, that is, so long as the dissociation curve remains as depicted in Fig. 85. Apart then from the possibility of oxygen

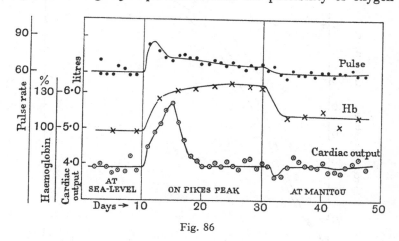

Fig. 86

secretion, the dissociation curve must alter, for at altitudes of over 24,000 ft. Greene (1932) records but little cyanosis. At about 14,000 ft. (Barcroft and others, 1923) such a change has been found but it has not been observed with certainty at lower altitudes.

Fig. 87 shows the dissociation curves of two normal individuals (Henderson, 1928) at about sea-level (at 40 mm. CO_2 pressure) determined by the van Slyke technique. The crosses and dots are those of two residents of Cerro de Pasco. Fig. 88 shows the same two normal dissociation curves, while in this case the dots are those of three members of our own party—A. V. Bock was a subject in both cases, his dissociation curve being the lower of the two normals.

This shift in the dissociation curve is of course prejudicial to the unloading of the blood and therefore prejudicial to the organism, it is a concession to the fact that oxygen must get into the haemoglobin before it can get out, but it tends to immobilise the organism. I have sometimes wondered whether it may not be the cause of the rather disappointing results obtained from the use of oxygen at high altitudes. If, at a moment's notice, the atmospheric pressure at sea-

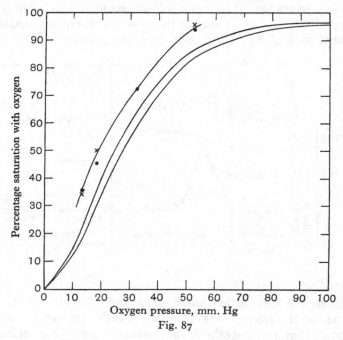

Fig. 87

level could be reduced to that of Everest, the immediate ascent of a mountain if given oxygen to breathe should present no difficulty, a fact which has been amply proved in chamber experiments but, in the case of persons acclimatised to the highest altitudes possible, oxygen seems less satisfactory, which is mysterious. The clue to the mystery may lie in the possibility that the dissociation curve, once having shifted to the left and immobilised the man, does not immediately return.

And may I digress a little further and ask why should it return immediately? I can only answer that we are in a realm in which there is any amount of work to be done. No dissociation curves have been determined at the appropriate altitudes, *i.e.* above 20,000 ft. on mountains. We have no more to go upon than the all too meagre information obtained at Cerro de Pasco, but the reader when looking at Fig. 87 cannot fail to recall Fig. 72 and to be struck

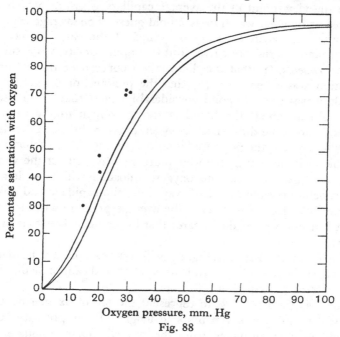

Fig. 88

with the fact that when man is replaced in the embryonic condition of oxygen want his dissociation curve tends to revert to its prenatal position. This fact was first pointed out by Anselmino and Hoffmann (1931). It will be remembered that the foetal haemoglobin is not identical with the maternal. If the condition of oxygen want pushes the curve to the left by causing some alteration in the alkalinity of the intra-corpuscular *milieu*, then the return to the normal might be fairly rapid. If, on the other hand, the dissociation curve can only return to its normal position as a result of breakdown of haemoglobin, then the process must be slow. The latter contingency would seem unlikely.

I should like to treat anaemic anoxia in the way in which I have just treated anoxic anoxia, but that seems impracticable. If the reader will bear with me I will explain the reason and it will not make a bad introduction to the problem. Anoxic anoxia was treated from the view-point of the average capillary pressure, which was arrived at by what was frankly a bold guess. The average capillary pressure was taken as being one-third of the way between the venous and arterial pressures. But that approximation was subject to two *caveats*, (1) that the dissociation curve over the region in question was steep, and (2) that the pressure of oxygen in the capillary was at every point considerably above that in the tissue. The rate at which the blood is losing oxygen in the capillary depends upon the difference of oxygen pressure between the tissue and the blood at the point under consideration. At the arterial end the blood is becoming reduced more rapidly than at the venous end, and therefore half the oxygen unloaded will have left the blood before the blood is half way along the capillary and, taking the capillary from end to end, the average pressure will be somewhat but not a great deal nearer that in the vein than that in the artery.

The following diagram (Fig. 89) will serve to illustrate the matter, though as given it is but a crude illustration and cannot be dignified by the term "calculation".*

Suppose the line $A-V$ to represent the fall of oxygen pressure along the capillary, seven units in length. If the pressure in the individual units be taken from the arterial to the venous end it would be as follows:

Unit	1	2	3	4	5	6	7	
Average pressure	76	63	54	47	42	38	36	Mean 51 %

The line was drawn quite at random and the average of the pressures (51) is very nearly equal to the venous pressure + one-third of the difference between arterial and venous pressures. The reader

---

* An actual calculation would be based among other things on the reciprocals of the figures.

can alter the factors involved, making the curve a little more or less
inflected and the pressure in the tissues a little higher than zero;
but he will find that so long as he keeps that pressure considerable
he will not greatly alter the result.

In the case of anaemia the whole picture is different. In our
consideration of the anoxic type we assumed that 35 per cent. of
the haemoglobin lost its oxygen in the circuit of the body, but what
if the blood contains only 35 or 40 per cent. of the normal quantity

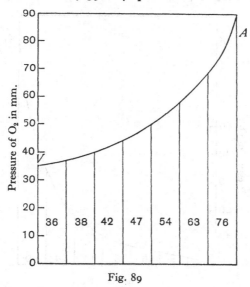

Fig. 89

of haemoglobin? If there were no compensatory mechanism the
blood must become almost completely reduced in order to supply
the tissues with their former quantity of oxygen; the curve of reduction in the capillary would become of the general type shown
in Fig. 90, and the oxygen pressures in the seven units would work
out something as follows:

Unit	1	2	3	4	5	6	7	
Oxygen pressure	50	24	14	8	6	4	2	Mean 15 %

The average 15 is very far from a third of the distance between
that at V_1 and that at A_1, which would be 29. In any case the low

pressures obtaining along the greater part of the capillary would be quite problematical.

The above figures afford no easy basis for calculating the average pressure in the capillary.

Two questions present themselves. What form of adaptation can take place? and how are we to treat of it? Fortunately the answer to these is the same. The first consideration is to obtain a reasonably high venous oxygen pressure. This will provide (1) a capillary

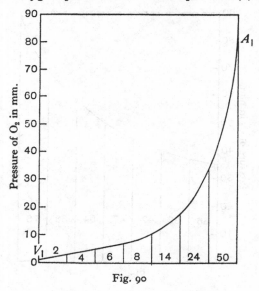

Fig. 90

pressure sufficient for the purposes of diffusion, and incidentally (2) a difference of pressure between the blood and the tissues large enough to justify our former method of guessing the average capillary pressure.

Now the essence of our difficulties is not really that there is too little haemoglobin in the blood, but that too little haemoglobin enters the capillary in a given time. We have assumed that the blood contained only 35 per cent. of the normal quantity of haemoglobin: if the circulation rate be speeded up in the ratio 100/35, or about three times, the normal quantity of haemoglobin will traverse the capillary in a given time, and if the normal quantity of oxygen were unloaded we should revert to curve *A–V*, Fig. 89. Regard this quickening of the circulation rate as Step 1.

Now we may return to our old method. Step 1 would consist in raising the mean capillary pressure from some quite nebulous point

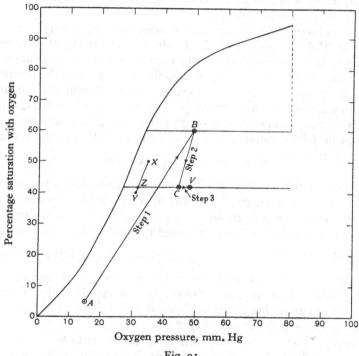

Fig. 91

which I shall put down on Fig. 91 as 15 mm. (Fig. 91, *A*) (because in our example it happened to work out so, but which might almost equally be 5 or 25) to a much more definite point one-third of the way between 34 and 80 mm., namely 49 mm. (Fig. 91, *B*).

The capillary conditions as regards gas diffusion would thus be restored to normal, but the body would be far from normal. The heart which was originally sending 5 litres of blood round the body per min. is now assumed to be sending 15, an amount corresponding to very considerable exercise. The further steps must be in the nature of arriving at a compromise between ideal conditions for capillary diffusion and excessive strain on the heart. The strain on the heart must be relieved, even though the conditions for capillary diffusion become less ideal.

The relevant data concerning anaemia are very scanty. In the case of T. J. F., quoted by L. J. Henderson (1928), the oxygen capacity of the haemoglobin of the blood was just below 7 volumes per cent. which would be approximately 35 per cent. of the normal. Arterial venous difference of oxygen combined with the haemoglobin was $6 \cdot 65 - 2 \cdot 88 = 3 \cdot 77$ c.c. per 100 of blood. If the blood were 97 per cent. saturated, the oxygen capacity of the haemoglobin would be $6 \cdot 86$ volumes per cent. and the saturation of the venous blood would be

$$2 \cdot 88 / 6 \cdot 86 \times 100 = 42 \text{ per cent.}$$

Let us plot this on Fig. 91. The position now would be, as the result of the second step, that the average capillary pressure of oxygen would drop from 49 to $44 \cdot 7$ mm. (Fig. 91, C), whilst the blood flow instead of being increased threefold would only be increased in the ratio $\dfrac{97 - 42}{35} = \dfrac{55}{35}$. If it started at 5 litres per min. it would now only be about 8. For a not very great sacrifice in the conditions of oxygen diffusion in the capillary the minute volume of circulating blood has been reduced from 15 at this end of Step 1 to 8 at the end of Step 2.

Step 3. There is one more alteration which tends to restore the oxygen capillary pressure, this time at the expense of the respiration, not the blood flow, namely an alteration in the dissociation curve. As pointed out by Henderson and by Litarczek and his co-workers (1929) in the case of man and, as has been shown in this laboratory by Litarczek and Stromberger in the case of experimental rabbits, anaemia is associated with a shift of the dissociation curve to the right. Taking Henderson's value for the shift as being one-sixth of the oxygen pressure, the curve at 50 per cent. saturation would have shifted from 30 to 35 mm. pressure and at 40 per cent. saturation from 26 to 30·7 taking up the position XY. The new venous pressure Z would become 32 mm. and the new capillary pressure V, 48. As the result of these three steps, an organism exists which at rest stands to have its tissues normally supplied with oxygen, but with its reserves seriously encroached upon. Its circulation rate is 8 litres per min. and not 5, whilst the increase of hydrogen-ion concentration to which the alteration in the dissociation curve is due will have pushed it to the point of breathlessness.

During exercise if the tissues are to maintain their supply of oxygen the blood flow must increase in the ratio of 8 : 5, and the breathlessness must become more marked. It is not necessary to follow the organism further; enough has been said to show that the relatively satisfactory position at rest is due to an integration of at least three factors. The condition of ischaemic anoxia, important as it is clinically, seems scarcely to merit detailed discussion, because that discussion would merely involve traversing the same ground as in the case of anaemia. The problem is the same, namely, that too little haemoglobin traverses the capillary in a given time. The reason is different—because the blood flow is too slow.

REFERENCES

ANSELMINO, K. J. and HOFFMANN, F. (1931). *Arch. Gynäk.* **147**, 69.
BARCROFT, J. (1920–1). *Nature*, **106**, 125, and Presidential address, Physiol. Section, Brit. Assoc., Cardiff.
BARCROFT, J., BINGER, C. A., BOCK, A. V., DOGGART, J. H., FORBES, H. S., HARROP, G., MEAKINS, J. C. and REDFIELD, A. C. (1923). *Phil. Trans. Roy. Soc.* B, **211**, 351.
GREENE, R. (1932). *J. Physiol.* **76**, 2P.
GROLLMAN, A. (1930). *Amer. J. Physiol.* **93**, 19.
HENDERSON, L. J. (1928). *Blood*, p. 267. New Haven.
LITARCZEK, G., AUBERT, H. and COSMULESCO, I. (1929). *C.R. Soc. Biol.* **101**, 220, 222.
LITARCZEK, G. and STROMBERGER, K. As yet unpublished.
PETERS, J. P. and VAN SLYKE, D. D. (1931). *Quantitative Clinical Chemistry*, p. 577. Baltimore.

CHAPTER X

THE "ALL-OR-NONE" RELATION

In the following pages the "all-or-none" relation will be considered with regard to certain tissues of the body. The order in which those tissues are cited bears no relation whatever to the chronology of the relevant discoveries. The order is physiological if it is anything: the tissues being placed rather with reference to the rapidity of their respective activities than with reference to the dates at which they became associated in men's minds with the all-or-none relation.

MEDULLATED NERVE

The proof that the all-or-none principle applies to medullated nerve was due mainly to Verworn (1913) and his pupils at Bonn and to Lucas and Adrian at Cambridge. The conception, of course, is that the disturbance which traverses a nerve fibre when the nerve is stimulated is a constant property of the fibre and not of the stimulus. The nerve fibre like the muscle fibre may alter its properties from time to time. Moreover, all nerve fibres are not the same: some are more easily stimulated than others. In a rough way one may liken the stimulation of a nerve fibre to the "touching off" of a trail of gunpowder. You may use a strong detonator or a weak one: but the gunpowder once alight, the subsequent progress of the combustion does not depend at all on the strength of the percussion which starts it, but only on the quantity and special properties of the gunpowder. Imagine that on an otherwise uniform trail of gunpowder there is a certain region in which the explosive has deteriorated (say it is damp); one of two things may happen when the flame reaches that portion. Either the gunpowder is so bad that the trail of fire may be entirely stopped or, if the deficiency is less marked, the flame may work its way past that region and will get to the sound trail again; when once it gets there (and that is the important point) it will propagate itself at the same rate and with the same intensity as if it had received no opposition on the way.

Now replace the trail of gunpowder by a nerve; subject a portion of the nerve to 5 per cent. alcohol in Ringer's solution and pass shocks into the nerve at regular intervals of about one shock every 20 seconds. Each shock produces a contraction of the muscle at the other end of the nerve—between the stimulating electrodes and the muscle is the alcohol. At first the muscular contractions are not altered; as time goes on, however, the alcohol injures the nerve increasingly and at a certain juncture, where the injury is sufficient, the contractions cease. Now if the size of the nerve impulses depends on the strength of the stimuli which set them in motion, we should expect to find that an increase in the strength of the shocks would make the muscle contract again, for a large disturbance would be more likely to pass through the injured region. But it is found that the strength of the stimulus makes no difference: a very strong shock is no more effective than one which would just produce a maximal contraction before the alcohol was applied. All disturbances, however they are set up, are extinguished at the same moment.

Again, if the disturbance gets through the impaired region at all it attains to the same characters in the subsequent length of sound nerve as it would have had if there had been no region of impaired conductivity for it to traverse. Adrian's (1912) original demonstration of this was to treat two lengths of the nerve with alcohol leaving a stretch of normal nerve between. The failure of conduction occurred no earlier than if only one of the lengths were treated with alcohol. On the other hand, if the two lengths were continuous, with no gap in between to allow the disturbance to recover, the failure of conduction occurred much earlier. The experiment seemed conclusive when it was originally made, but it rested on the idea then current that the disturbance becomes progressively smaller and smaller as it passes through the impaired region. Kato (1924, 1926) has shown good reason to doubt the possibility of such decremental conduction. He regards the region of the nerve treated with alcohol as exhibiting the all-or-none relation equally with the sound nerve, but the relation as being quantitatively different. To return to the classical analogy of the gunpowder, the treated nerve is a trail of gunpowder like the untreated but of worse gunpowder. The length of nerve treated with alcohol makes no difference to the time taken to suspend conduction, provided that the length is great enough to overcome the effects of diffusion gradients at the margin of the chamber. Adrian's finding that two lengths together were more potent than two lengths with a normal stretch between was probably due to such diffusion. But there is no doubt that the disturbance does return to its original size when it leaves an impaired region, for both Kato and Davis, Forbes, Brunswick and Hopkins (1926) have shown that the electrical response does so however much it may have been reduced in the treated area, and Tsai (1931) has confirmed this recently with a single fibre instead of an entire nerve.

In the particular case of mammalian sensory nerves there was the additional complication that the reflex evoked by stimulation of a sensory nerve appeared to be graded. The gradation took place both in respect to the height and duration of the contraction when elicited reflexly (Sherrington, 1921). A possible reconciliation between Sherrington's observation and the all-or-none relation was suggested by Forbes and Gregg as early as 1915: namely "that the wide variation in the reflex response to single stimuli is due to the setting up of more than one impulse when the stimulus is strong".

Adrian and Forbes in 1922 took up the point. They proved not only the validity of the all-or-none relation for sensory nerves in the mammal, but also the hypothesis that a single strong stimulus might release a series of impulses up the sensory fibre, and so produce a more powerful effect on the central nervous system than would have been elicited by a weaker stimulus initiating a single sensory impulse. Clearly the acceptance of the all-or-none relation as true for sensory nerves raises a number of points about the nature of sensation. The gradation of sensations is a common experience. One cup of tea is warmer than another, one light is brighter than another, one sound louder than another, and so forth. What mechanism accounts for such a gradation? Again, what is the essential difference between one sensation and another? As regards graded contraction of a skeletal muscle we have considered two factors—(1) the number of fibres which are being stimulated, and (2) the number of impulses which reaches each in a given time. In the case of sense organs the frequency of impulses set up is a very important factor.

While different sense organs react somewhat differently, the muscle spindle may be taken as an example. This organ has been

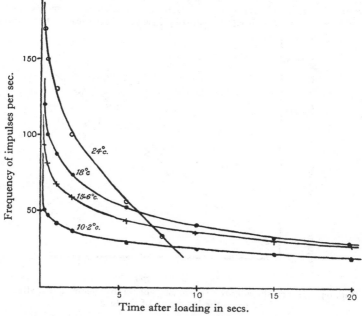

Fig. 92. Graph to show the decline in frequency of the response from a single end-organ after a load of 2 gm. is hung on a thread from the tendon. (Matthews.)

studied by Adrian and Zotterman (1926) and more recently by Matthews (1931). Matthews has made a quantitative study of the frequency of the impulses which pass up the fibre from a single muscle spindle when it is stretched. When the muscle is extended suddenly (*i.e.* in about 1/5 sec.) by a weight, the impulses reach a maximal frequency in about that time. The frequency rather quickly falls off, but so long as the tension lasts there is a rhythmic discharge of impulses up the fibre (Fig. 92). This tendency for the frequency to diminish, though the tension remains, is characteristic and important. In some sense organs the frequency declines much more rapidly. Indeed according to Adrian (1930) the phenomena shown by a muscle spindle and a hair respectively are at opposite ends of the scale:

When a hair is bent, impulses pass up the sensory fibre only during the actual movement of the hair and there is no further discharge when the movement ceases, although the hair is maintained in an abnormal position. When a muscle is stretched the sensory discharge reaches its maximum

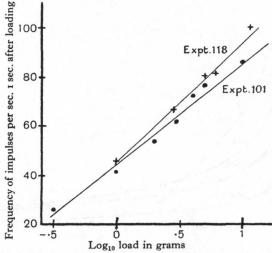

Fig. 93. Graphs showing the relationship of the frequency 1 sec. after loading to the logarithm of the load.

frequency during the period of extension, but if this is afterwards maintained at a fixed value the discharge continues at a lower frequency which declines very gradually. Here as in the hair the maximum frequency depends on the rate of stretching; after the movement has ceased the frequency depends upon the amount of stretch and the time since the beginning of stimulation.

And so we come to the next point, the relation between the frequency and the degree of stretching at a given time after the commencement of the stimulation. This again has been worked out by Matthews (1931); it appears to be a logarithmic one, the frequency varying with the logarithm of the weight suspended from the muscle as is shown in Fig. 93.

From the physiological point of view the significance of the continuance of the stream of impulses over the whole period of deformation of the muscle spindle lies in the proprioceptive function of the spindle; the central nervous system requires to be informed not only that the muscle has become stretched but that it remains in a condition of tension, and of the degree of tension that is maintained. Moreover, the logarithmic relation between the frequency of the discharge from the nerve ending and the strength of the stimulus is not without its significance. It seems to bring us very near to the Weber-Fechner Law and seems to suggest that the just perceptible difference between two weights varies directly with the number of impulses which pass up the nerve. I say "seems to bring us nearer the Weber-Fechner Law" because though the similarity between Matthews' statement and the Law itself is evident, the foundations on which they rest are less similar. Matthews' statement has to do with a single fibre—the Weber-Fechner Law appears to be a relation derived from the study of a portion only of the curve obtained by the integration of the properties of a whole population of fibres. Thus if there were a great number of fibres each of which gave a curve similar to that of Matthews, the result obtained from the whole nerve would be a curve of a sigmoid type. The Weber-Fechner Law is the expression of the central portion AB of this integration (Fig. 94).

In the above experiments stretching was adopted as the appropriate stimulus for the nerve endings in muscle. The legitimacy of such a course is perhaps open to doubt on first principles, for we might expect shortening rather than lengthening to be the property of the muscle which the nervous system is most anxious to appraise. As a matter of fact the experimental evidence for the adoption of stretching rather than contraction is complete.

When a muscle which is in a state of slight tension contracts, the occasional rhythmic impulses which had been ascending the sensory fibres cease, and during the actual period of contraction there is "silence" followed by a volley of impulses at the commencement of relaxation (Fig. 95). The importance of this observation in the appraisement of what takes place during rhythmic processes such as respiration is evident enough.

So far we have considered the frequency of the rhythm as its only variable property, perhaps giving the reader the impression that as a corollary of the all-or-none relation, alteration of frequency of afferent impulses is the only means by which the central nervous system can be informed of alterations in the intensity of the stimulus which causes them. But it must be remembered that a strong stimulus will affect more sense organs than a weak, so that

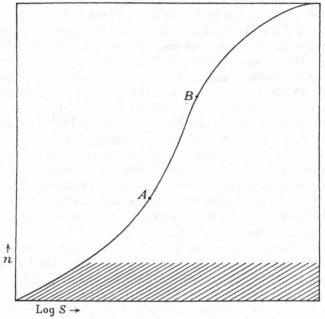

Fig. 94. See text, p. 235.

the number of sensory nerve fibres in action will vary as well as the impulse frequency in each nerve fibre. There are however some other possibilities. Even if we confine ourselves to the consideration of a single fibre, inspection of Fig. 96 shows that all the records of the impulses are not all of the same time relation. The variation, L, of a rapid series is more acute than its successors M, N, etc., which appear to be conducted more slowly than L; the fully developed variation is only to be found again towards the end

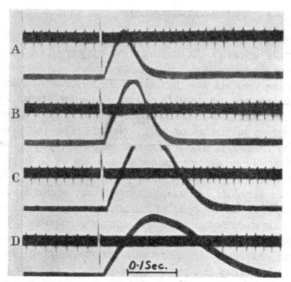

Fig. 95. The effect of fatigue on the pause in the response. Isotonic twitch superimposed on a steady tension of 1 gm.; each record was taken between 1 and 2 sec. after the muscle was loaded. Temperature 14° C. *A*, muscle rested; *B*, after 50 twitches; *C*, after 100 twitches; *D*, after 200 twitches. The line in *D* represents 0·1 sec. for all.

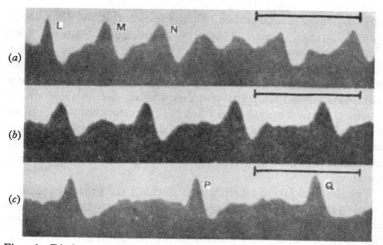

Fig. 96. Discharge produced by sudden application of a 20 gm. load, (*a*) was taken coincident with loading, *L* being the first impulse of the series; (*b*) 2 sec. later; (*c*) 5 sec. later. The line in each represents 0·01 sec.

of the record, *e.g.* P and Q when the impulses are separated by considerable intervals. At first sight we appear to be faced with a violation of the all-or-none relation, but this is not so—at least it is not necessarily so. It was implicit in the all-or-none relation from the first that the tissue should be in the same condition throughout. Otherwise Bowditch's staircase effect (as will be shown later) would be a violation of the all-or-none relation. The fact that N, M, etc., differ from L is due to the rate at which they follow it and one another. They do not follow so quickly as to be within the absolute refractory period (if so they would never materialise), but outside the absolute refractory period there seems to be a short interval of time—"the relative refractory period"—within which the stimulus does not produce its complete response.

It is perhaps going outside the strict limits of a discussion of the all-or-none relation to enquire whether the response in any two fibres is necessarily the same. This enquiry however cannot be altogether avoided if the all-or-none relation is under consideration in its setting, namely how a succession of all-or-none responses can produce diverse sensations in the central nervous system. Moreover, even if it were quite irrelevant, such is the beauty of Gasser and Erlanger's (1927) work, that I would probably be unable to pass it by without a reference. Their thesis is that the response in any two fibres is, in general form, the same; but the rate at which it is conducted varies from fibre to fibre. The larger the fibre the more rapid is the rate of conduction. Of sensory fibres those connected with muscle would appear to be the largest, and therefore if an impulse is sent from the periphery, along the nerve of mixed sensation, the first impulses to reach the posterior root will be those of muscular sense. Also it has been shown by Matthews (1929) that the excitation wave set up by stimulation of end-organs in the skin is different in shape from that set up by pulling on a muscle.

[The demonstration which makes perhaps the greatest personal appeal to me, considering the place at which I started physiology, is the following: the lingual nerve centrally to its crossing of Wharton's duct carries large sensory fibres which come from the tongue and small fibres which supply the submaxillary gland. The two separate near the crossing of Wharton's duct by the lingual nerve to take different courses peripherally. It is therefore possible to pick up what is happening in each separately. Travelling along the same length of fibre an impulse set going in the mixed nerve will reach a given point on the lingual nerve (peripheral to Wharton's duct) before it will reach a corresponding point on the chorda tympani. The difference in the rate of conduction corresponds to the sectional area of the fibre concerned.]

However, as Adrian (1930) points out, it is not enough that the fibres corresponding to different sensations should conduct at different rates if, as Gasser and Erlanger suggest, the specific form of impulse plays a part in determining the paths open to them in the grey matter of the spinal cord, the specific forms must be preserved to the final central terminations of the fibres, which are far smaller than the peripheral fibres leading to them, and the form of the impulses depends upon the size of the fibres at any point. To make Gasser and Erlanger's theory complete the relative differences of fibres from different types of nerve ending would have to be preserved in their central terminations.

So far we have discussed the all-or-none relation in the sensory mechanism especially with reference to muscular sense and we have said something about tactile hairs. Before leaving the subject, however, something must be said about vision, the consideration of which carries us a step further in "placing" the all-or-none relation as a feature in physiological function.

Much of the information has already been obtained by the study of single end-organs, a muscle spindle, a Pacinian corpuscle, a tactile hair and so forth—together with the attached nerve fibre. At present there is no technique so refined as to give the response elicited by the stimulation of a single rod, or a single cone in the vertebrate retina, and if there were the "attached fibre" does not penetrate beyond the level of an adjacent layer in the retina.

It is otherwise, however, with the crustacean eye, in which it is possible to stimulate a single facet and to investigate the response set up in a single fibre which leads from it. This has been done in Bronk's laboratory by Hartline and Graham (1932). I quote a letter in which Dr Hartline kindly gave me an account of what he is doing:

By good luck and fortunate choice of material we have been able to record impulses in single fibres of the optic nerve upon stimulation of the eye by light. Moreover, the nerve fibres in the preparation we have used come directly from the photosensitive cells, without any intervening synapses or ganglion cells, so that the records are, I believe, strictly comparable with those obtained from end organs of tension, touch, pressure, etc. The animal used is the horse-shoe crab, *Limulus polyphemus*. The lateral eye of this crab contains about 300 ommatidia, each about 0·1 mm. in diameter, and each furnished with a sensitive element composed of a central rhabdom about which are grouped 14–16 retinula cells. From each retinula cell a single nerve fibre runs uninterruptedly in the optic nerve to the central ganglion. In the larger animals the length of optic nerve between eye and central ganglion may be as much as 10 cm. It is very easily dissected out and the eye and optic nerve when excised will survive in a moist chamber 10–12 hours. You see, this optic nerve is not comparable to the optic nerve of the vertebrates, but is truly a sensory nerve, rather than a sort of peripheral brain tract. The fibres would correspond probably to rod or cone fibres in the vertebrate eye (or perhaps to the dendrites of the bipolar cells—the junction between retinula cell and nerve fibre might conceivably have physiological properties of a synapse). It is very fortunate that this optic nerve possesses only a very delicate connective tissue sheath, and it is quite easy to split the nerve up into a number of fine bundles. Using glass needles and working under a microscope it is possible to obtain bundles which contain scarcely more than a dozen or so fibres. Now in the adult animal I believe the eye must undergo a certain amount of degeneration—the shell is not moulted and the cornea of the eye may become very badly eroded, so that a fair percentage of ommatidia are no longer functional. This means that the optic nerve contains a number of inactive fibres which "dilute" the active ones. It frequently happens, therefore, that a small bundle dissected off the nerve contains only three or four, or perhaps only one active fibre. This condition is easily recognised—the discharge of impulses in a bundle containing only one fibre consists of regularly spaced impulses of perfectly uniform height—just like the single fibre discharge from a muscle tension receptor. It is easy to show that this single fibre discharge is obtained upon stimulation of one particular ommatidium—the surface of the eye is explored by means of a small spot of light, and impulses are obtained only from a small region, which examination shows to be occupied by an ommatidium. When several active fibres are by chance present in the nerve bundle, it is frequently possible to recognise the impulses belonging to the several fibres as they are apt to differ in size and in amount of diphasic recording. The impulses in each fibre, moreover, constitute a regular series, which differs in frequency from the discharge of the other fibres and which shows a different threshold of illumination. In one case, where only two active fibres were present, the impulses in one were almost twice the size of those in the other, and it was possible to obtain either discharge alone by confining the stimulating light to one or the other of two adjacent ommatidia. Ordinarily, the restriction of the area illuminated is not alone sufficient to give single fibre responses which can be recorded from the entire optic nerve—one must rely on the good fortune of the dissection to single out only one of the 14–16 fibres coming from one ommatidium.

The discharge in a single fibre of the optic nerve closely resembles that obtained from other sense end organs. In response to continuous illumination of the eye impulses are discharged at a frequency which initially is quite high (the highest obtained has been 130 per sec.), but which falls, rapidly at first and then more slowly, approaching a steady value which is about one-tenth the initial value. The discharge persists as long as light continues to shine on the eye, and stops abruptly when the light is turned off (there is no "off effect" in this eye). The effect of intensity is very striking: (1) the latent period (between onset of illumination and the first impulse) is shorter the higher the intensity; its reciprocal is linear with $\log I$; (2) the frequency of discharge, both the initial value and the final steady one, is higher the more intense the light. The plot of frequency against $\log I$ is S-shaped, so that within a moderate range of intensity the frequency is linear with $\log I$—the well-known Weber-Fechner relation, which has also been found by Matthews in the tension receptor. The approximate nature of the relation is well known in visual physiology. At low intensities the discharge becomes irregular, as in other sense cells, and may even stop. Just above the threshold it consists of a single impulse. The response of the completely dark adapted receptor to light of high intensity shows a striking feature—after the initial maximum (0·3 sec. after the onset of illumination) the discharge is interrupted for about 0·2 sec., although the light shines steadily. After this brief silent period the discharge is resumed at a much lower frequency and is maintained without further interruption at a constant frequency for the duration of illumination. The silent period is absent if the eye has previously been adapted to a bright light and returned to darkness for only a few minutes. The response to even the highest intensity is then an unbroken series of impulses, starting at a high frequency which falls rapidly to a steady value. Coming from this material this is quite interesting, for I think if it were observed in a vertebrate optic nerve one would almost certainly ascribe it to the synapses or ganglion cells. As a matter of fact, Matthews has noticed a similar phenomenon in tension receptors—it is not nearly so marked, and Matthews is uncertain whether it may not be due to failure to apply the stimulus absolutely uniformly. Here there can be no such doubt.

Work on this preparation is interesting for two reasons: first, it may help to unravel some of the points of general interest with respect to receptor mechanism. It is very much easier technically to control a light stimulus than almost any other kind of stimulating agent. The advantage of this is shown in connection with the "silent period". Then, too, it is very difficult to study the relation between intensity and duration of short periods of stimulation, using any other stimulus than light.

The second point of interest in connection with this work is its relation to the physiology of vision. Single fibre records from a vertebrate optic nerve, if they could be obtained, would be doubly interesting, now that one has a general idea of the type of discharge originated by the receptor cells themselves. The S-shaped relation between $\log I$ and visual effect, and the general shape of the dark-adaptation curves are both well known in visual physiology—it is quite interesting to see them both represented in the behaviour of a single photoreceptor unit. Of particular interest is the wide range over which intensity is effective in producing responses of varying magnitude. This range may be as great as 1 to 1,000,000—it is remarkable to find that the single photoreceptor cell covers it functionally.

Let us turn to the electrical variations in the optic nerve of the eel. These have been studied by Adrian and Rachel Matthews (1927, 1928). Between the optic nerve and the actual rods and cones there is at least one synapse, and therefore in studying the variations in that nerve we are studying the sensory impulses as modified by the intervening synapse or synapses.

Suppose aa' (Fig. 97) consists of a number of receptor organs the fibres of which end in the synaptic layer bb' and that C_1, C_2, C_3, C_4 represent fibres in the optic nerve.

Though it is impossible to obtain the response from a single element in the vertebrate retina, Adrian and Rachel Matthews have left us not entirely uninformed as to what we might expect of such a response. The record of potential given by the optic nerve of the eel when a small area of the retina is illuminated consists of a jumble of peaks or serrations. It is unlike the skyline of the Alps, but resembles that of the Andes (or what I have seen of the Andes) in that the individual peaks resemble one another very closely and that there are "lots and lots" of them. The resemblance is so close as to convince the authors that they were dealing with a number of all-or-none responses, spaced at random but following one another at intervals so short as to make these effects overlap to some extent.

The effect of increasing the intensity of light when the field of illumination is small is to increase the number of impulses which pass up the optic nerve. That also is the result of increasing the area of the field, if the intensity of illumination be maintained constant*. But the relation of these two increments is different from what might have been expected. Consider first increase of intensity. Thus, if a single fibre were stimulated with light of varying intensities, one would expect, from Matthews' work and that of

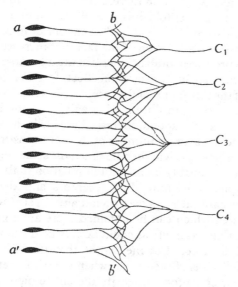

Fig. 97. aa' receptor organs; bb' synaptic layer; C_1, C_2, C_3, C_4 fibres in optic nerve.

Hartline and Graham, the number of responses per second to increase not in proportion to the intensity but to the logarithm. On the other hand, if the intensity were kept uniform and the area increased on the assumption that the number of fibres was correspondingly increased, one would expect that the number of responses in the optic nerve would increase proportionately to the number of fibres stimulated, that is, to the area. Thus a given

* Another effect which does not immediately concern us here is that as either the intensity of the light or the size of the area stimulated increases, the latency of the retina shortens so that impulses appear in the optic nerve sooner after the illumination.

increment in the total quantity of light would produce a greater number of responses if it were distributed so as to increase the illuminated area than if it were distributed so as to increase the intensity. But something quite different takes place—the increment in the number of responses is the same in either case. It is not multiplied by ten by a tenfold increase in the area and therefore one must suppose that a tenfold increase of area stimulated does not involve a stimulation of tenfold in the number of fibres. But there is another possible explanation of the fact that increase in illumination or in area produces similar changes in the discharge of impulses; both increments may have a common effect, for as Hecht (1927) has pointed out, if the thresholds of the receptors in a small area vary over a wide range, increase in illumination will make the less sensitive receptors in that area respond, increased area will enable the stimulus to take effect on a larger number of the most sensitive receptors. The effect in either case will then be much the same; more receptors are stimulated. Probably increased illumination both increases the response frequency of individual receptors and the number in action. In any case the responses do not alter in size and even with the smallest area which can be obtained there are a great many of them, say ten times as many as could be expected from the stimulation of a single fibre. Let me repeat that the above statement applies to small areas of the retina; when the whole retina is stimulated with light of uniform intensity the entire picture alters. The waves in the record become large and smooth; it is not to be supposed, however, that there is any violation of the all-or-none relation. The explanation appears to be that the ganglion cells placed between the individual end-organs and the fibres of the optic nerve when stimulated in sufficient numbers set up a concerted discharge with a definite rhythm, and that each large wave represents the aggregate of the individual all-or-none responses of a large number of fibres whose cells are discharging more or less simultaneously.

STRIATED MUSCLE

Though the all-or-none relation had been discovered for cardiac muscle in 1871, it was not till 1909 that Lucas thrilled the physiological world in general, and the laboratory in which he worked in particular, by his demonstration of the obedience of striated muscle and nerve to the all-or-none principle. The key to this demonstration was the discovery of a nerve in the frog which had only ten fibres. That key, so Lucas generously says, was put in his hand by Anderson. The *cutaneus dorsi* nerve of the frog contains not more than a dozen fibres, commonly less than that number. Of these one at least is sensory, and two others are fine medullated fibres which do not supply skeletal muscle. After these deductions have been made eight or nine fibres remain which, when stimulated, cause the fibres of the *cutaneus dorsi* muscle to contract. The number of muscle fibres involved is, of course, much more numerous, for near the point of entry of the nerve into the muscle, the nerve fibres undergo a series of divisions so that the original eight axons in the nerve control between 150 and 200 muscle fibres. Each nerve fibre therefore with its twenty or so attached muscle fibres forms a unit.

The nerve was stimulated by a current of known strength and the height of the muscular contraction was recorded. The strength of the stimuli ranged between one so weak as just not to give a contraction of the muscle and one just so strong as to give a maximal contraction. Between these strengths the relation of the strength of the stimulus to the height of the contraction was plotted as a graph. The essential point which Lucas observed was that this graph was not a smooth curve but a staircase. True the steps are high at the bottom and small at the top, but it is a staircase all the same.

The interpretation put by Lucas on his diagram (Fig. 98) is as follows.

Presumably each step in the increase of contraction means that another nerve fibre (or group of fibres of like excitability) has been excited, and that the muscle fibres to which that nerve fibre runs have consequently contracted. It follows that in each muscle fibre the contraction is always maximal regardless of stimulus which excites the nerve fibre.

The skeletal muscle cell of amphibia therefore resembles the cardiac striated muscle cell in the property of "all-or-none" contraction. The difference which renders it possible to obtain "submaximal" contractions from a whole skeletal muscle but not from a whole heart is not a difference in the functional capabilities of the two types of cell; it depends upon the fact that cardiac muscle cells are connected one with another, whereas skeletal muscle cells are isolated by their sarcolemma. The "submaximal" contraction of a skeletal muscle is the maximal contraction of less than all its fibres.

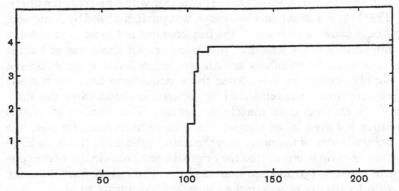

Fig. 98. Diagram constructed to show the relation between the magnitude of the contraction and the strength of the exciting current on a true scale. Abscissae represent the strength of the exciting current, that which just excited the nerve being taken as 100. Ordinates measure the height of the contraction, the increase due to the favourable effect of previous contractions being neglected. (Keith Lucas.)

The widespread interest aroused by this experiment, the way in which it focussed attention on the all-or-none relation, is all the more noteworthy from the fact that it is not really conclusive as far as the muscle fibre is concerned, and that Lucas himself had made much more conclusive experiments four years earlier. When he showed that a muscle supplied by ten nerve fibres gave not more than ten steps in its contraction, he proved that each motor unit—each nerve fibre and its attached muscle fibres—must follow the all-or-none relation, but he did not show that the muscle fibres themselves would do so. The relation might have been due to the nerve fibre, the nerve ending or the muscle fibre, and if the nerve impulse cannot vary with the stimulus, one could scarcely expect the muscle response to the nerve impulse to do so. But in 1905 he had made much the same experiment on curarised muscle fibres stimulated directly and not by way of the nerve. He had cut through all but ten or so of the muscle fibres and had then recorded the contraction of these ten with an optical lever arrangement. When the stimulus was increased smoothly, the contraction increased by sudden steps and, as in the later experiments, the number of steps was never greater than the number of intact fibres. The results are quite clean cut, but the technical difficulty of the experiment may have made Lucas prefer the later version; taken together, the two experiments constitute a proof of the relation both in muscle fibre and in nerve fibre, but of the two, the earlier is the more conclusive.

It was confirmed by Pratt (1917) who first alone and later with the collaboration of Eisenberger (1919) planned experiments which would have pleased the heart of Lucas. They went further than he did by reducing the element stimulated to a single fibre—this was accomplished by the use of a "pore electrode". The pore electrode consisted of a glass tube drawn off to a very fine closed end. The end was then polished until the polished surface reached the lumen making a hole of less size than the diameter of a red blood corpuscle. The lumen was filled with "agar" and placed on a fibre of the frog's sartorius by the use of a micro-dissection apparatus or "micro-manipulator".

This electrode supplied unipolar stimulation to the fibre on which it rested. The other novel point about Pratt's researches was the method of registration of the movements of the muscle. Anything of the nature of a lever was too crude for a single fibre; a fine globule of mercury was therefore placed on the fibre. A beam of light reflected from the surface of the drop gave a graphic record of the movements of the fibre. The fibre was not, however, removed from the muscle. Pratt and Eisenberger's researches supported entirely Lucas's conception. The final paper of this series was published in 1919. By this time the all-or-none conception had derived an interest from sources far outside the superficial realm of the physiology of muscular contraction. Plank had disclosed the quantum theory. The obvious analogy between the quantum of energy in the world of material structure and the all-or-none discharge in that of muscular contraction, made an immediate appeal to certain minds in spite of the great disparity in the quantities of energy respectively involved. The result seemed a complete vindication of the all-or-none relation between the stimulus and the contraction in individual fibres of the intact sartorius muscle of the frog, and all seemed to go happily for that relationship until a few years ago when the frog's retro-lingual membrane came into fashion as a preparation for the study of single muscle fibres. The membrane consists of a single layer of parallel fibres, each separated from its fellow by a sufficient stretch of connective tissue. Fischl and Kahn (1928) found what they thought was a graded response in the single fibres, but Hintner (1930) and Pratt (1930) failed to confirm their work. In 1930, however, Gelfan found undoubted evidence of graded contractions when the fibre was stimulated by a pore electrode with a hole not more than 5μ in diameter. Gelfan used Pratt's method of recording, by photographing the beam of light reflected from a mercury globule in the muscle.

It was some time before Gelfan was able to convince physiologists of the accuracy of his results. They seemed to amount to a complete reversal of the all-or-none relation and yet that relation seemed based on other evidence of the strongest kind. But Gelfan's later work with Gerard (1930) and with Bishop (1932) shows how the conflict can be resolved. The micro-electrode can produce two distinct effects according to the strength of the stimulus, (1) a graded contraction localised more or less to the region of the stimulating current, not conducted down the fibre and not accompanied by an electric response, and (2) an all-or-none contraction accompanied by the usual electric response and propagated throughout the fibre. With larger electrodes only the second effect is produced.

Thus the graded contraction of a single muscle fibre is an undoubted fact, but it can only be obtained with certain types of electrode applied in certain ways and there is no reason to suppose that it occurs in the normal development of the contraction of the fibre when it responds in the natural way to impulses reaching it from the nerve. Thus the modern work, though revealing some phenomena only hinted at by Lucas, has detracted in no direction from his conception and has in many ways strengthened it.

We appear to have followed the all-or-none relation as far along academic lines as is desirable. The particular thrill that it imparts is an academic one; like, I suppose, the quantum theory, it spreads abroad the impression that it is drawing the worker nearer to the fundamental basis of things. But we must leave its academic side. It is not for me to discuss whether striated muscle is really all-or-none, or merely appears to be so because the gradation between stimulus and response is too delicate to be measured by methods so far devised. I am alive to the issue, but incompetent to discuss it. My business is with the mechanism of the ordinary graded contractions which constitute the natural movements of the body. The architecture of such movements is clear. A muscle contracts by one-half the amount of which it is capable, not because all its fibres undergo a sub-maximal contraction, but because the number of fibres in contraction is at any one time fewer than if the muscle were reduced to its minimal length. Each contracted fibre appears to undergo the full degree of shortening of which it is capable at the moment. The obvious demonstration of this essential feature in the architecture of muscular function is due to Adrian and Bronk (1929). An electrode consisting of a hypodermic needle which contains an insulated metal core is thrust into the arm, and if this is done in a fortunate manner the electrode will pick up the current from but a few fibres. This may be amplified and either registered photographically or turned into sound and announced by a "loud speaker". The latter mode of demonstration is perhaps less accurate but more arresting. So long as the muscle is at rest only occasional "clicks" are heard, each like its predecessor. As the muscle is thrown gradually into an ever-increasing degree of activity the "clicks" become more numerous, but they do not become louder. The general effect is that of a machine-gun firing at an increasing speed, but the noise made by the explosion of each cartridge does not alter.

THE HEART

The all-or-none principle appears to have been first enunciated by Bowditch (1871) in words of which the following translation is given by Bayliss (1915): "An induction shock produces a contraction or fails to do so according to its strength: if it does so at all it produces the greatest contraction that can be produced by any strength of stimulus in the condition of the muscle at the time". The preparation which Bowditch used was the perfused apex of the frog's ventricle of which, be it noted, the author made histological preparations to insure that nervous elements were absent from the tissue. The only active substance present therefore was cardiac muscle.

That was in 1871. The fact was confirmed by Luciani (1873) and Kronecker and Stirling (1874). So at least Luciani puts the matter. Stirling (1888), however, claims the discovery for his colleague and himself: "The heart either contracts or it does not contract, and when it contracts the result is always a maximal contraction" (Kronecker and Stirling). That there should be any doubt about the origin of the discovery is the more surprising because all the workers in question were working at the same time, I think, in Ludwig's laboratory. As often occurs in larger laboratories in which there are numbers of researchers, a new conception may have been in the air, and it is not impossible that the real originator of the all-or-none law was Ludwig himself.

At all events, as presented in Leipzig the principle did not hold beyond the domain of cardiac muscle and, indeed, up to 1909 obedience to the all-or-none law was held to be a property which was specific for cardiac muscle, which differentiated it from other forms of muscle and, indeed, from other forms of living animal matter.

Bowditch was clear on two accessory points. Firstly that the maximal contraction which the frog's heart would give differed at different times, and secondly the liminal stimulus necessary to evoke a contraction was not always the same.

A familiar demonstration of the fact that the heart is capable of a greater contraction at some times than at others is the "staircase effect".

As regards the strength of stimulus necessary to evoke a contraction Bowditch classified such into fallible and infallible stimuli. The former, as the name suggests, sometimes produced a reaction; the latter, stronger, of course, always evoked a beat.

It is perhaps worth pausing to consider how Bowditch's discovery appears in the light of the work of more recent researchers. If the reader will look back to the commencement of this chapter he will discover behind Adrian's argument a fact not as yet mentioned, namely that the disturbance suffers no decrement in passing down the fibre. It does not tend to fade out.

In its twentieth century guise the all-or-none principle appears as being something which is a property of the activity of a unit— a single muscle fibre or a single nerve fibre—and it is not unreasonably a little surprising that forty or so years ago it was originally discovered as constituting a property not of a single cell but of an extensive tissue, namely the heart.

Lucas! when you wrote the following words what exactly had you in your mind? "The skeletal muscle cell of amphibia therefore resembles the cardiac striated muscle cell in the property of all-or-none contraction. The difference which renders it possible to obtain maximal contractions from a whole skeletal muscle but not from a whole heart is not a difference in the functional capabilities of the two types of cell; it depends upon the fact that the cardiac muscle cells are connected with one another, whereas skeletal muscle cells are isolated by their sarcolemma". The meaning of your words is clear enough, but behind that clear meaning do you not raise some point on which your own opinion would have been invaluable? Does your observation not boil down to this: that the whole heart of the frog is a unit in the sense that one skeletal muscle fibre or one nerve fibre is a unit? and if so, is not what you term the cell, i.e. the portion of cardiac muscle presided over by one nucleus, but partially separated from its fellow by a diaphragm of material which stains with silver nitrate? is not this portion of cardiac muscle comparable merely to the part of a skeletal muscle fibre which may be referred to a single nucleus, although there is no obvious diaphragm between that part and its fellow?

And may we suppose, indeed is it necessary to suppose, that the disturbance goes over the whole heart without decrement?

If we accept the assumptions (1) that the heart of the frog is functionally a single but complicated cell, and (2) that the disturbance goes over it without decrement (or at all events that the decrement is always the same), we can raise no difficulty to its conformity with the all-or-none principle. Where then stands the heart of the mammal? Does the all-or-none principle apply to this mammalian heart? Is it a single cell? Does physiological continuity exist between its innumerable strands of fibres? Does the disturbance suffer no decrement as it passes along them?

There are some experiments, as was pointed out to me by Dale, which give a *prima facie* impression that the mammalian heart does not conform to the all-or-none principle—that it behaves more like a skeletal muscle than like a single fibre. The experiments in question are those of Cullis and Tribe (1913) on the effect of severing the auriculo-ventricular bundle. The first point to grasp is that in the mammalian heart muscarine has no direct effect upon the ventricle. If the bundle is completely severed, the ventricle beats with a rhythm of its own and the administration of muscarine, though it profoundly affects the auricular beat, produces no effect whatever on that of the ventricle (Fig. 99).

Now to turn to the effect of muscarine when the bundle is intact. This effect is shown in Fig. 100. The beats of the ventricle follow those of the auricle not only in frequency but also, and this is the important point, in strength. If it could be supposed that muscarine depressed the ventricular muscle the matter would present no difficulty, but we have just seen that the drug does not do so. In the old days before the work of Lucas and Adrian we would have supposed, simply, that after muscarine a less "disturbance" than normal travelled down the muscarinised A.V. bundle and met with a correspondingly less response on the part of the muscular fibres of the heart. That conception is clearly a direct negation of the all-or-none principle. Discarding this last explanation it is still open to us to analyse another, namely (1) that the muscarine depresses the A.V. bundle as alcohol would a nerve, (2) that it depresses some strands of the bundle to the point at which the impulse does not pass, but some strands only, (3) that the portions of the heart supplied by these strands do not give any contraction—whilst the portions supplied by the conducting strands give a maximal end contraction. Such a conception would entail no violation of the all-or-none relation as such, but it would place the mammalian

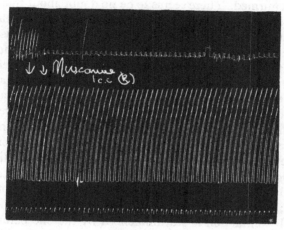

Fig. 99. Cat. Perfused heart. Injection of muscarine. A.V. bundle cut. In upper the auricles beat only once after injection; small movements due to pull of ventricle in auricle lever.

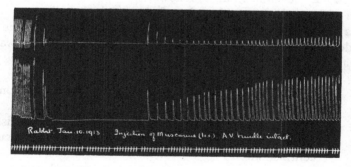

Fig. 100. Perfused heart of rabbit. Injection of muscarine. A.V. bundle intact.

heart in the same category with skeletal muscle, as being a tissue capable of grading its contraction, by regulating the number of contracting units in a different category from the heart of the frog, which *as a tissue* obeys the all-or-none relation.

Miss Alison Dale (1930) has repeated the work of Miss Cullis and Miss Tribe, considering it from the angle of its bearing on the all-or-none relation. The observation illustrated in Figs. 99 and 100 she has confirmed. There seems to be no doubt about that, but she presents an interpretation of it which does not violate the all-or-none relation. This interpretation is as follows: When muscarine or arecoline is exhibited to the heart, true it acts only on the auricle, true the ventricular contractions become reduced in height, but they also take place at less frequent intervals. The familiar "staircase" effect illustrates the fact that if two contractions take place, the one immediately after the other, the second is of greater amplitude than the first. The excess is due in some way not understood to the fact that during the second contraction the heart is still under the influence of the preceding one. That is the basis of Miss Dale's explanation; the ventricular contractions under arecoline are so far separated in time that each takes place independently of its predecessor; ordinarily in the case of the rabbit's heart that is not the case, each contraction is still influenced by the one before it, and therefore marks higher on the drum than it otherwise would do.

We are left then with a conception of the mammalian ventricle which may be summed up as follows:

(1) That it is, from this point of view, a single cell.

(2) That it obeys the all-or-none relation and that there is no evidence of decrement in the conduction of the contractile wave. As a corollary:

   (a) The impulse which arrives along any one twig of Purkinje tissue would produce a complete contraction of the whole ventricle.

   (b) The point in having so rich a branching of Purkinje tissue is to insure approximate synchronism in the ventricular contraction.

(3) That the mammalian ventricle, unlike striped muscle, is unable to grade its activity by altering the number of units which contract at any one time. Considerable variations in its activity are, however, possible, and the beat can alter (1) in frequency, and (2) in strength, the alteration in strength being dependent on some change in the conditions under which it is acting.

CONCLUSION

The natural transition would be from cardiac to unstriped muscle, but about the latter tissue we know almost nothing in this connection. It is clear that unstriped muscle may contract to very different extents. If the whole of the frog's heart must be regarded functionally as a single fibre, the same view must be taken with even more certainty about a mass of unstriped muscle. The conception of a graded contraction being due to complete contraction of some fibres whilst others are completely relaxed is a very difficult one when the fibres as in unstriped muscle appear to form a continuum. Striped muscle is capable of graded contraction because, though the all-or-none relation holds, the fibres are separate units; cardiac is incapable of graded contraction because the all-or-none law holds and the fibres are not separate units; the deduction with regard to unstriped muscle would be that as it is capable of graded contraction, as the fibres are not separate units, the all-or-none relation does not hold. Moreover, it is clear that the contraction in unstriped muscle in many cases dies out, in which it does not resemble the impulse governed by the all-or-none relation in either striated muscle or nerve. On the other hand, unlike the contracture, it does move though slowly from the immediate seat of stimulation. At present, however, and pending further work we are forced to conclude that the principle under discussion does not apply to unstriped muscle.

The same is true of gland cells. We do not know the essential nature of glandular secretion, but though it is evident, and of this we shall speak in another connection, that active glands contain areas which appear not to have secreted recently, there is not real evidence for the all-or-none relation between secretion and the electrical stimulus which brings it about. Years ago when I worked on the salivary glands Hardy often suggested that one factor in secretion might possibly be of the nature of electrical endosmose. That view has been favourably discussed since by Loeb and others. On any such hypothesis one might expect the water which passed through the cell to vary with the difference of potential between its two surfaces, and there seems to be no reason why that difference

256   THE "ALL-OR-NONE" RELATION

of potential should not vary with the stimulus. I suppose if one
tried to visualise it one might do so thus, that with a graded stimulus
applied to the chorda a greater or less number of all-or-none responses reaches the nerve endings and produces a greater or less
quantity of something like pilocarpine, to which the cells would
react by setting up a greater or less electric potential. That however
is perhaps going too far into the region of speculation.

Till unstriped muscle and secreting cells are shown to exhibit the
all-or-none relation it is not possible to regard that relation as one
which is observed by living tissue generally—rather it is a form of
specialisation, a price—as I have heard from those about me who
study the subject—paid by tissues which react very rapidly for
extreme rapidity and mathematical certainty of response.

If the view just stated is correct, the all-or-none relation is nothing basal but is something which has been evolved from more
primitive forms of tissue which do not possess it: observe how I am
not asking you to spell evolution with a very large E.

REFERENCES

ADRIAN, E. D. (1912). *J. Physiol.* **45**, 389.
—— (1914). *Ibid.* **47**, 460.
—— (1930). *Physiol. Rev.* **10**, 336.
ADRIAN, E. D. and BRONK, D. W. (1929). *J. Physiol.* **67**, 119.
ADRIAN, E. D. and FORBES, A. (1922). *Ibid.* **56**, 301.
ADRIAN, E. D. and MATTHEWS, R. (1927). *Ibid.* **63**, 378; **64**, 78.
—— —— (1928). *Ibid.* **65**, 273.
ADRIAN, E. D. and ZOTTERMAN, Y. (1926). *Ibid.* **61**, 151, 465.
BAYLISS, W. M. (1915). *Principles of General Physiology*, p. 452. London.
BOWDITCH, H. P. (1871). *Ber. Sächs. Ges.* **23**, 652.
CULLIS, W. C. and TRIBE, E. M. (1913). *J. Physiol.* **46**, 141.
DALE, A. S. (1930). *Ibid.* **70**, 455.
DAVIS, H., FORBES, A., BRUNSWICK, D. and HOPKINS, A. McH. (1926).
   Amer. J. Physiol. **76**, 448.
FISCHL, E. and KAHN, R. H. (1928). *Pflügers Arch.* **219**, 33.
FORBES, A. and GREGG, A. (1915). *Ibid.* **39**, 211.
GASSER, H. S. and ERLANGER, J. (1927). *Ibid.* **80**, 522.
GELFAN, S. (1930). *Ibid.* **93**, 1.
GELFAN, S. and GERARD, R. W. (1930). *Ibid.* **95**, 412.
GELFAN, S. and BISHOP, G. H. (1932). *Amer. J. Physiol.* **101**, 678.

HARTLINE, H. K. and GRAHAM, C. H. (1932). *Proc. Soc. Exp. Biol. Med.* **29**, 613.
HECHT, S. (1927). *J. Gen. Physiol.* **11**, 255.
HINTNER, H. (1930). *Pflügers Arch.* **224**, 608.
KATO, G. (1924). *The Theory of Decrementless Conduction in Narcotised Region of Nerve.* Tokyo.
—— (1926). *The Further Studies on Decrementless Conduction.* Tokyo.
KRONECKER, H. and STIRLING, W. (1874). *Beitr. Anat. Physiol.* and *A Text-book of Human Physiology.* Landois, translated by Stirling, 3rd ed. London, 1888.
LUCAS, K. (1905). *J. Physiol.* **33**, 125.
—— (1909). *Ibid.* **38**, 113.
LUCIANI, L. (1873). *Luciani's Human Physiology*, **1**, 318. London, 1911.
MATTHEWS, B. H. C. (1929). *J. Physiol.* **67**, 169.
—— (1931). *Ibid.* **71**, 64.
PRATT, F. H. (1917). *Amer. J. Physiol.* **44**, 517.
—— (1930). *Ibid.* **93**, 9.
PRATT, F. H. and EISENBERGER, J. P. (1919). *Ibid.* **49**, 1.
SHERRINGTON, C. S. (1921). *Proc. Roy. Soc.* B, **92**, 245.
TSAI, C. (1931). *J. Physiol.* **73**, 382.
VERWORN, M. (1913). *Irritability.* Yale Univ. Press.

CHAPTER XI

UNITS

Several circumstances within the last fifteen years have conspired to focus attention upon the subject of "units". In the first place the establishment of the all-or-none relation in respect to the contraction of the fibres of skeletal muscle has reduced the conception of the shortening muscle to a calculation of the number of fibres in contraction at any one time as opposed to the degree of contraction of any one fibre. In the second place, the researches of Krogh on the capillary circulation have drawn attention to the regulation of the circulation by the number of open and shut capillaries, as opposed to, or rather as supplementary to, the degree of latency of the arterioles. Again the researches of A. N. Richards and his pupils have emphasised the possible influence on renal secretion of the number of glomeruli which are in action.

Then there are much older strata of thought. As far back as the days when Dr Anderson instructed me in the basal principles of the sympathetic system, stress was laid upon the multiplication of units as a means of operating a very great number of peripheral structures from a relatively small number of cells in the central nervous system. At each relay the number of fibres involved in the transmission of the impulse from a single cell in the brain is multiplied. The general scheme applies, of course, not only to the sympathetic system but to the innervation of the skeletal muscles and, I suppose, the peripheral sense organs.

And in addition there was the Great War which moulded the minds of the participants into the habit of thinking in terms of the number of units thrown into action, so that at the time when Krogh, Richards, Sherrington and Adrian stressed the subject in the economy of the body, the soil was, as it were, fully prepared for the reception of the seed.

The instances of unitary function which I have cited are not all of the same type: roughly speaking they may be divided into two, those of which the cell is the unit, and those of which the organ is the unit, if I may use the word "cell" to describe a muscle fibre and "organ" to describe the tubule of the kidney.

SKELETAL MUSCLE

Skeletal muscle seems to be the organ which functions most obviously by the mobilisation of units. Keith Lucas (1909) was the first to show that a muscle composed of, say, two hundred fibres and supplied by ten nerve fibres can be made to contract in ten steps according to the number of nerve fibres stimulated. The muscle therefore is made up functionally of ten units, each consisting of about twenty fibres.

In the mammalian muscle which he has studied, Sherrington (1931) estimates the "motor unit" as consisting of more numerous fibres:

Soleus (cat) has some 30,000 muscle-fibres and the long extensor of the digits some 40,000 (Clark, 1930); but nervous coordination, as Hughlings Jackson said, is "based on the anterior horn-cell". That cell by its motor nerve-fibre innervates a whole packet of muscle-fibres, 150 and more. This packet of muscle-fibres together with its motor nerve-fibre constitutes what may be called a "motor unit". It is into these packets that the reflex fractionates its muscle. They give numbers not unmanageable— *soleus* (cat) consists of some 200, *extensor longus digitorum* 330, medial *gastrocnemius* 450 (Eccles and Sherrington, 1930).

Each such motor unit has a physiology of its own. It must all contract at once because the nerve fibre which supplies the motor unit observes the all-or-none relation as also do the muscle fibres themselves. Each motor unit can exert a definite tension and, as some units within the same muscle contain more fibres than do others, the tension exerted by different units is different.

Such motor units yield a contraction-wave of some 2·5 gm., and even of 9 gm. contraction-tension. The contraction waves overlap to form a tetanus. This tetanus when once the mechanical fusion of the component waves is *complete* yields in its turn a tension value fixed, in so far that no increase of strength or rate of stimulus will change it further. This full tetanic tension value for the average unit is 10 gm. in *soleus* (cat) and reaches 35 gm. in *gastrocnemius*.

Indeed so fixed would the properties of a motor unit appear to be that the question must arise, why have all these fibres? Why not

have a single fibre of the size of, and exerting the power of the whole
"packet" of 130 or so which make up the motor unit? The answer
I suppose is a question of nutrition. When the muscle is in full and
prolonged contraction, in the steady contracted state the fibres require to be fed with oxygen, etc., at a certain rate; for the efficient
conditions of respiration no part of the muscle fibre must be more
than a certain distance from the capillary and the combined muscle
fibres must have a sufficient surface. This suggestion is borne out
by the fact that the fibres of more intense metabolism in the warm-
blooded animals are smaller than their analogues among the lower
vertebrates. There would appear to be two sorts of units conditioned by different consideration, the *fibre* the size of which
depends upon the possibilities of nutrition, and the *motor unit* the
size of which depends upon functional activity. But what considerations of functional activity decide the number of fibres in a
motor unit? Is it simply that, say, 130 muscle fibres is a convenient number for a nerve cell to supply and that, say, 1300 fibres
would be for some reason beyond the capacity of the nerve cell?
If I have read the work of Sherrington and his colleagues correctly,
the size of the unit is regulated by conditions of a quite different
character, namely that operated by the various permutations and
combinations of afferent stimuli; the motor units should be of such
a size as to give the most suitably graded contractions as the reflex
response to afferent impulses.

CAPILLARIES

The researches of Krogh (1922) have led to the conception of
"open" and "shut" capillaries, so that passages in his book leave
the reader with the impression that the quantity of oxygen which
can pass to a tissue depends upon the number of open capillaries
which supply it. He writes:

My own first contribution to the problem of capillary contractility...
was undertaken to test the hypothesis of a regulation of the supply of blood
to muscles through the opening and closing of individual capillaries.
I found it possible to observe at least the superficial capillaries of muscles
both in the frog and in mammals through a binocular microscope, using
strong reflected light as a source of illumination. Resting muscles
observed in this way are usually quite pale and the microscope reveals

only a few capillaries at fairly regular intervals. These capillaries are so narrow that red blood corpuscles can pass through only at a slow rate and with a change of form from the ordinary flat discs to elongated sausages. When the muscle under observation is stimulated to contractions a large number of capillaries become visible and dilated, and the rate of circulation through them is greatly increased. When the stimulus has lasted only a few seconds the circulation returns in some minutes to the resting state, the capillaries become narrower and most of them are emptied completely, while a small number remain open....

In resting muscles of the frog the average distance between open capillaries, observed simultaneously through the microscope, was estimated at $200–800\mu$, but after contraction this could be reduced to 70 or 60μ.

So much for the fact of "open" capillaries, and now, as regards their relation to the respiration of the muscle, Table XXV may be quoted.

Table XXV

	Number of capillaries per mm.² cross section	O_2 consumption per min. vols. % of the tissue
Rest	31 85 270	0·5 1·0 3·0
Massage	1400	5·0
Work	2500	10·0
Maximum circulation	3000	15·0

Indeed the doctrine of open and shut capillaries has become so familiar that the reader may wonder why I did not choose the capillaries first for consideration. The reason is that the case does not appear to be so clear as that of muscle. In it the fibres either are contracted to their full or are relaxed, there is no half-way house; but the case of the capillaries is different, for it is clear from Krogh's description that there are degrees of openness. To go no further than the quotation that I have given above, Krogh says a large number of capillaries become visible *and dilated*.

I need not labour the point further than to refer the reader to Fig. 101 which shows capillaries of very various calibres, the calibre of any one of them being subject to considerable variation. There is no suggestion in Krogh's work that the Rouget's cell which he regards as controlling the calibre of the capillary contracts on the all-or-none principle; in fact he states quite the opposite. Some constituent factor in the capillary wall possesses, like smooth muscle, a definite tonus, or "posture", to use the terminology of Sherrington (1920)—the calibre of the capillary is on the whole determined by the state of contraction of the factor in question.

Though capillary "units" therefore form a leading feature in the regulation of the peripheral blood flow, the peripheral resistance, even so far as the capillaries alone are concerned, is not simply an

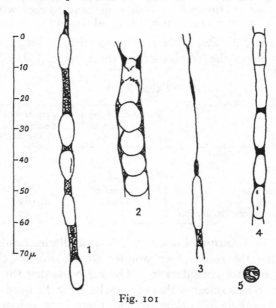

Fig. 101

effect of so many capillaries open and so many closed, it is a mixed effect involving not only the number of capillaries open but the degree of their openness.

GLOMERULI

A third sphere in which the principle of units has come into considerable prominence of late years is that of the kidney. The conception is due to the brilliant work of A. N. Richards and his associates.

The whole topic is so well treated in Richards's Harvey Lecture (1922) that I cannot do better than quote him, albeit at some length.

In our earliest experiments the variability of our preparations was striking. In some preparations as many as eight glomeruli could be counted in a field of 2 mm. diameter; in others only three or four in the whole kidney, in so far as it was accessible to inspection. Our frogs were pithed, and if two or three drops of blood were lost in this operation the number of glomeruli to be found in the kidney was small. If, however, such a frog were immersed in a saline bath, or if his abdominal cavity were filled with isotonic salt solution, the number of visibly active glomeruli increased.

Acting on the suggestion which this fact afforded, we have made a series of counts of the glomeruli which show active circulation under varied conditions.

Bits of silk thread were laid transversely across the surface of the kidney at approximately equal distances of about 2 mm. (the diameter of our low-power field). From five to eight fields were thus separated for ease in counting. A bit of cover-slip was lightly laid over these, both to prevent displacement of threads and to avoid surface glare.

I shall cite figures to show alterations in the number of glomeruli in which blood was flowing before, during and after the introduction of various substances to be mentioned.

1. *Isotonic salt solution.* As has been mentioned, salt solution is absorbed from the open abdominal cavity of the frog, and as a result circulation improves if it has been lessened from hemorrhage. These figures illustrate: soon after preparation five fields showed 5 active, 8 inactive, total 13 glomeruli: 30 minutes after salt solution had been introduced into the belly the same fields showed 28 active, 0 inactive, total 28 glomeruli. (By "active" glomeruli we mean those showing active circulation.)

2. *Injection of blood.* 0·5 c.c. of whole blood was taken from the aorta of one frog and immediately injected into the anterior abdominal vein of a frog, whose glomeruli had been counted. Before injection, active glomeruli 10, inactive 27, total 37; 5 minutes after injection, active glomeruli 39, inactive 9, total 48.

3. *Injection of isotonic salt solution.* 0·5 c.c. of 0·6 per cent. Before, 41 active, 9 inactive, total 50; 10 minutes later, 54 active, 8 inactive, total 62; 20 minutes later, 44 active, 6 inactive, total 50.

4. Then *urea.* 0·1 c.c. of 20 per cent. After, 65 active, 2 inactive, total 67.

5. *Caffein.* Seven fields counted: before injection, 81 active, 11 inactive, total 92; 0·1 c.c. of 2 per cent. *caffein,* 5 minutes later, 104 active, 0 inactive, total 104.

6. *Glucose.* 0·1 c.c. of 10 per cent. glucose. Before injection, 31 active, 12 inactive, total 43. Followed by progressive increase in number of active until 35 minutes after, 62 active, 3 inactive, total 65.

7. *Hypertonic sodium sulphate* (5 per cent.). 0·1 c.c. Before, 6 active, 0 inactive, total 6; after 13 minutes, 51 active, 0 inactive, total 51.

8. *Adrenalin.* Constrictor dose, 0·1 c.c. 1 to 100,000. Three fields: before injection, 49 active; immediately after, 12 active; 7 minutes later, 48 active.

9. *Pituitrin.* Constrictor dose, 0·1 c.c. 1 to 10 dilution of pituitrin "S". Before, 14 active, 5 inactive, total 19; after, 0 active, 16 inactive, total 16; later, 5 active, 12 inactive, total 17.

These and many similar observations have led to the conclusion that even under the most favorable of operative conditions, *i.e.* with the least loss of blood, all the glomeruli of the kidney of the frog do not receive blood simultaneously. Conditions which depress the circulation, such as blood loss or destruction of the cord, or agencies which constrict blood vessels in the kidney, such as constrictor doses of adrenalin or pituitrin lessen the number of glomeruli which receive blood. Plethora, absorption or injection of isotonic salt solution, hypertonic NaCl, hypertonic sodium sulphate, urea, glucose and caffein—all are capable of impressively increasing the number of glomeruli which receive blood at one time.

Not only is the number of glomeruli showing active circulation altered by the agencies which I have mentioned, but also the number of capillary loops within a single glomerulus which take part in the capillary blood flow. Earlier in this section I referred to glomeruli of two rather widely different aspects in so far as blood flow through them is concerned; one in which narrow rapidly flowing currents of blood indicate a complex network of tortuous channels; others in which one or two loops only are visibly filled with blood and in these blood usually flows more slowly and in a wider stream.

The dilator agencies, urea, caffein, etc., mentioned above have the power of transforming a glomerulus of the latter type into one of the former.

So much for the frog. The situation is not so clear with regard to the mammal; as presented by Richards (1925) it is as follows:

Kanolkhar, in London, injected hemoglobin solutions intravenously into animals, removed the kidney and in thick sections of the cortex counted the number of glomeruli which showed traces of hemoglobin and of those which did not, he came to the conclusion that events similar to those which we had described in the frog's kidney took place in the mammalian kidney. Dr Hayman and Dr Starr, in our laboratory, have carried through a series of experiments similar to these in design. They used Nelson's method for estimating the number of glomeruli. This consists in staining the glomerular tufts by passing a solution of Janus Green B through the renal vessels, dissecting the cortex away from the medulla, and counting the stained tufts in a small weighed section of the cortex. Knowing the weight of the whole cortex, the total glomerular count for the whole kidney is easily ascertainable. After convincing themselves of the reliability of the method, they found that the number of glomeruli in the two kidneys in a normal rabbit is approximately the same. Then the following procedure was adopted. From the anesthetized rabbit suitably prepared, one kidney was excised without hemorrhage, and laid aside for subsequent enumeration of the total number of glomeruli in it. The condition to be studied was then established in the animal (that is, injection of diuretic, of adrenalin, and so forth), a solution of Janus Green B injected into the aorta above the remaining kidney, and within 20 seconds of the end of the injection this kidney was excised. Obviously the dye could reach only such glomeruli as were receiving blood at the time of injection and during succeeding seconds.

Glomerular count of the intravitally stained kidney compared with the total glomerular count of the other showed the fraction of glomeruli which were receiving blood during the presence of the dye in the circulation. A long series of experiments was made, the results of which are convincing. They showed that from 50 to 85 per cent. of the total glomerular equipment of the rabbit receive blood under control conditions. If a diuretic, such as caffein or NaCl, was given from 95 to 100 per cent. of glomeruli receive blood. If graded doses of adrenalin were given, or if vasoconstriction were produced by inhalation of CO_2 or if the splanchnic nerve were stimulated, the number of glomeruli receiving blood could be diminished at will to 40, 30, 20, 10 or 5 per cent. of the total glomeruli present. These experiments appear to me to furnish adequate proof that the observations made on the frog are acceptable for the interpretation of events which take place in the mammalian organ.

It is much to be hoped that further developments may be possible along the lines which Richards has so skilfully laid down. Speaking as one who once worked on the kidney, it has long seemed to me that progress along the old lines had rather nearly ceased, and failing some quite new development it was difficult to see what could be done. Richards's work seems to offer the prospect of a new field more fertile than that which had been so completely worked out. Meanwhile I can only point out that the glomerular tufts form a system even further removed from a rigid one of open and shut units than is the capillary, for an open glomerular tuft consists of capillaries, as Richards has shown, some of which may be open and some shut and the open capillary may be of greater or lesser calibre according to circumstances.

266   UNITS

REFERENCES

KROGH, A. (1922). *The Anatomy and Physiology of Capillaries*, p. 37. New Haven.

LUCAS, K. (1909). *J. Physiol.* **38**, 113.

RICHARDS, A. N. (1922). *Amer. J. Med. Sci.* **163**, 1. Harvey Lecture, 1921–2.

—— (1925). *Ibid.* **170**, 781. Mary Scott Newbolt Lecture.

SHERRINGTON, C. S. (1920). *The Integrative Action of the Nervous System.* Yale.

—— (1931). *Brain*, **54**, 1.

CHAPTER XII

THE PRINCIPLE OF ANTAGONISM

One of the most striking features in the architecture of function is the antagonism of certain nerves. The burden of the present chapter will be to follow up some of these antagonisms, and the result to conclude that, whether viewed from the angle of structure or of function, they conform to no common type.

In the last decade of last century much was heard of antagonistic nerves. As the present volume makes no pretence of being a history it would be completely redundant to discuss the various theories which have been held with regard to them.

It was pointed out that organs as different as the heart, the blood vessels, the salivary glands, the pupil and the retractor penis had a double innervation, that the sympathetic quickened the pulse while the vagus slowed it, that the chorda tympani dilated the vessels of the sub-maxillary gland while the sympathetic constricted them, that the third nerve constricted the pupil while the sympathetic dilated it, that the *nervus erigens* produced contraction of the retractor penis whilst the hypogastric produced extension.

The theory of antagonistic action as understood by Gaskell (1900), who in my student days used to fascinate us with it, may be set out in his own words:

The conclusion to which I came in 1881 (*Phil. Trans.* London, **173**, 1029) was that the vagus nerve was the anabolic nerve of the heart, and that inhibition or relaxation of muscular tissue was a sign of anabolism in that tissue, just as contraction of the tissue is a sign of katabolism in that tissue. Subsequently I have put forward the proposition that all muscular tissues are probably supplied with anabolic and katabolic nerves, the one causing relaxation of the tissue and diminution in its contractions, the other causing contraction of the tissue.

Such double nerve supply has been proved for many muscular tissues, both striped and unstriped. Thus the muscles of the vascular system are supplied by vaso-dilator as well as vaso-constrictor nerves; the unstriped muscles of the bladder and other visceral muscles are supplied with two opposite kinds of nerves: among these the recent observations of Langley and Anderson (*J. Physiol.* Cambridge and London, 1895, **19**, 71) have shown a specially good example in the case of the retractor penis, supplied as it is with motor nerves from the lumbar region and inhibitory nerves from the sacral region; the adductor muscles of *Anodon* are supplied with two nerves of opposite function, according to Pawlow (*Arch. f. d. ges. Physiol.* Bonn, 1885, **37**, 6) and Heidenhain. And the most suggestive case of all, in connection with the question of the action of antagonistic striated muscles in vertebrates—the adductor and abductor muscles of the claw of the crayfish are, according to Biedermann's researches, supplied with two nerves, of which the one is motor and the other inhibitory.

It may, I think, be fairly said that this view, that the action of the vagus nerve is to promote the anabolic or assimilatory processes in cardiac muscle, is steadily gaining ground.

To the instances cited above another may be added which Gaskell often discussed, that of the salivary glands, to which the chorda tympani nerves were regarded as being anabolic and the sympathetic katabolic.

The theory of anabolic and katabolic nerves seems to have died a natural death. For its disappearance there is probably more than one reason. In the first place it purported to be a chemical theory of the living process; as such, the chemist might rightly demand that it should be stated more definitely and in the language of known chemical reactions. Gaskell, indeed, in conversation used to give the theory a more specific character—he regarded anabolism as especially linked up with the absorption of oxygen and katabolism with the production of CO_2. Since those days around 1900 much water has flowed under the bridges. Gaskell's views no doubt were based on the old inogen theory of muscular contraction; according to it inogen—a material rich in "intra-molecular oxygen"—was stored in the muscle, and the contraction was due to the explosion of the stored inogen. The series of researches commenced by Fletcher and continued by Hopkins, Hill, Meyerhof, Embden and, lately, the Eggletons and Lundsgaard, have given us a very different theory of muscular contraction. Hill's researches seem to show also that the chemistry of nervous excitation is of the same type. The modern view does not present metabolism as a process which divides itself into anabolism and katabolism. It is natural, therefore, that the theory of antagonistic nerves should cease to be of interest from the old standpoint, namely that of its being the key to the essential process of living material.

Yet if the antagonism of nervous action has been dethroned from its former high estate, the fact of its existence remains. Nay more, for if the chemical background of Gaskell's views proved to be illusory, his statement of them proved to be prophetic. Let me go back to the quotation which I have already made: "And the most suggestive case of all, in connection with the question of the action of antagonistic muscles in vertebrates—the adductor and abductor muscles of the claw of the crayfish are, according to Biedermann's researches, supplied with two nerves of which the one is motor and the other inhibitory". It remained for Gaskell's own pupil Sherrington to enrich his generation with discrete knowledge of the physiology of the joints, but surely the passage quoted above shows some glimmering on the part of Gaskell of the direction in which that knowledge was to be found. To this might be added much that Magnus has taught us about the posture of the joints.

The theory of anabolic and katabolic nerves has gone; the point I wish to stress is that it was a theory of the cell metabolism and that the essence of cell metabolism was supposed to be the antagonism of these two processes each ruled by a separate nerve. Has any such fundamental theory replaced that of anabolic and katabolic nerves, and if so, does it involve any principle of antagonism? The answer is I think in the negative, but let us examine some examples: the nerves which supply the salivary glands, the blood vessels and other tubes, the retractor penis, the pupil and the striated muscles which control the eye ball and the joints.

Of these I will consider the salivary glands first rather with the object of disposing of them, for in truth it is not easy to see that even superficially they present an example of antagonistic action. I speak of the cells, not the vessels.

THE SALIVARY GLAND

It is a curious irony that the sub-maxillary gland came to be considered as a site for the study of cell metabolism reduced to its simplest terms. The study of all the salivary glands is complicated, because it is impossible when the nerves are stimulated to dissociate the cellular from the vascular effects. Not until someone devises a scheme of irrigating the gland with a blood supply which is at once constant, adequate and independent of the nervous impulses which reach the cells, will satisfactory knowledge be forthcoming with regard to the action of the parasympathetic and sympathetic nerves, but of all the salivary glands the sub-maxillary is the least simple. Situated more or less where the inside of the body meets the outside, it is innervated by the nerves of each, the sympathetics which supply the sweat glands and the parasympathetics which supply the glands of the stomach. These meet in the sub-maxillary, and what then? Have the sympathetics any direct effect on the cell at all? Some deny it. Langley once said to me: "Everybody who works at the salivary glands believes at some stage that the sympathetic has no effect on the cells and ultimately gives up that belief". Certainly the more recent workers (Babkin and McLarren (1927)) hold that the sympathetic does play in some direct way on the cells. And their views agree in this: that the effects of the two nerves on the gland, though different, are not antagonistic. It is not certain even that any single cell is acted on by both nerves. We may therefore cross the salivary glands off the list and proceed to consider unstriped muscle.

UNSTRIPED MUSCLE

In unstriped muscle there appear to be four conditions which are more worthy of consideration:

    (1) Complete contraction.   (3) Peripheral tone.
    (2) Central tone.   (4) Complete relaxation.

    The position as it seems to me with regard to the smooth muscle of the blood vessels is that the pressor nerve is responsible for central tone up to and including complete contraction; cut the pressor nerve and peripheral tone remains. Cut the depressor nerve at that stage and nothing happens. Stimulate the depressor nerve and peripheral tone is removed. To take an example from the sub-maxillary gland of the dog (weight about 10 gm.):

Table XXVI

	Blood flow c.c. per min.
Chorda tympani stimulated (complete relaxation)	10
Chorda and sympathetic cut (peripheral tone) ...	2
Sympathetic intact (central tone)	1
Sympathetic stimulated (full constriction) ...	0·1

The chorda tympani and the sympathetic are antagonistic in this sense, that the sympathetic secures the extreme contraction of the fibre and the chorda tympani the extreme dilatation. But we may ask further questions:

(1) Do they act in opposite ways on the same mechanism in the muscle fibre or do they act on different mechanisms?

(2) Do the pressor and depressor nerves combine to subserve a single physiological function or are they functionally distinct?

(3) To what extent are they distributed?

(1) To address ourselves to the first question:

(a) We must provide some more exact proof that peripheral tone though capable of being abolished by nervous influences is a condition of the cell and not the effect of nerves upon it.

(b) That it is a function of the *life* of the cell.

(c) That there are two distinct processes in the muscle which regulate its length.

(a) The view was once rather attractive that muscle in a state of tone was necessarily supplied by a nervous plexus and subject to the influence of the processes taking place in the plexus. The plexus was really the seat of the tone. Most of the work on this subject has been done on the gut. It is not necessary to rehearse the evidence in favour of the existence of such plexuses from Auerbach's plexus downwards; the difficulty of workers to obtain a preparation of unstriped muscle free from all nervous elements has been very great. And indeed the view that the contractions of the muscular coats of the small intestine depend upon their government by Auerbach's plexus formerly had all the authority of Magnus's support. Magnus found

that it was possible to separate the outer longitudinal layer of muscle (with covering of serous membrane) from the inner circular muscle with the attached mucous membrane and submucosa. After this procedure the longitudinal muscle which had been stripped off, was found to retain its power of rhythmic movement but the circular muscle had for ever lost its power. By histological examination he found that Auerbach's plexus which is situated between the circular and longitudinal layers of muscle adheres almost exclusively to the latter. Thus plenty of nerve cells could be discovered in the stripped off longitudinal muscle but few or none in the remaining circular portion.

The above account of Magnus's experiment is given by Gunn and Underhill (1914) in the prelude to their own consideration of the subject. Not only were they unable to repeat Magnus's experiment, but their work provided reasons for taking the opposite view, namely that the tone and rhythm of the intestinal muscle could be maintained without any assistance from the nervous element of Auerbach's plexus.

They found that, in preparations which had been kept so long (116 hours) in the cold that the ganglion cells might be supposed to have lost their efficacy, the muscle still contracted rhythmically and still was capable of alterations in tone produced by drugs, and that the power of rhythmic contraction disappeared after about the same number of hours in cold storage, as did the influence of barium. They also obtained tonic contraction with adrenaline and relaxation with pilocarpine in preparations which they believed to have no nervous elements. In the above account of Gunn and Underhill's work I may seem to have accepted rhythmicity as proof of the capability for tone. Clearly, unless or until reasons can be given for regarding rhythm and tone as separate manifestations of the same phenomenon, it is right that in discussing the evidence rhythm should not be accepted blindfold as tone, but quite clearly in many of Gunn and Underhill's experiments tone was present in muscles which, so far as the authors could judge, were free from ganglion cells. More recently the ground has been covered again by Gasser (1926); the particular merit of his research is perhaps scarcely brought out with sufficient emphasis in his paper, and he will perhaps excuse me if I say a word on the subject. It was the extreme pains that were expended on the histological scrutiny of the muscles. The modest statement made by Gasser on this subject is: "A fourth strip gave the record shown in Fig. 8. This strip was through the kindness of Prof. Magnus taken for examination by Dr Van Esveld. It was stained by him with his own haematoxylin and the stain was sufficiently successful to reveal ganglion cells if present; however none were found". The muscle in this experiment was sent for inspection to the master in such matters, and it may be taken for granted that if any method could have revealed the presence of nerve cells it would have been Van Esveld's, and if any eye could have seen them, it would have been Magnus's.

Fig. 8 in Gasser's paper shows that the strip of the circular muscle of the intestine, on which such care was lavished, on treatment with acetyl choline gained not only in rhythm but in tone. Other strips of unstriped muscle, apparently equally free from nervous elements, suffered reduction of tone with adrenaline, atropine, nicotine, etc., which at least shows that they had previously been in the condition which we have described as peripheral tone.

It would be interesting to repeat the results on a preparation of muscle from the arterial wall if such could be found free from nerve cells.

(b) That peripheral tone is a manifestation of the life of the fibre may, I think, fairly be conceded from the fact that either want of oxygen or treatment with small quantities of cyanide abolishes it; at all events this is the case in blood vessels, as we shall see, and in the muscle of the spleen, as shown by de Boer and Carroll (1924).

The picture of peripheral tone at which we have arrived is thus that it depends upon two things, firstly the aerobic life of the muscle fibre, and secondly its exposure to some sort of chemical environment.

(c) The actual process or processes involved in the contraction of unstriped muscle have provided a subject of research for a group of workers in recent years: Evans, Brocklehurst and L. E. Bayliss, Hill and Winton. Perhaps this work may be discussed rather from the most recent backwards. Winton (1930) has shown in the retractor penis, which is devoid of nerve cells and in which stretching produces no nervous response, that there are two types of metabolic activity which he calls respectively (α) postural, and (β) phasic.

(α) It is typical of postural activity that a very small weight will gradually stretch the muscle by a very large amount if allowed to exert its influence for a considerable time. Thus stretched the muscle is very viscous. It is difficult to increase or decrease its length rapidly. Such activity makes very little demand on the metabolism of the muscle, and therefore the oxygen consumption of the muscle may be the same whether the muscle is long or short, provided it is not undergoing rhythmic contractions.

(β) On the other hand the phasic contractile element is associated with rather large differences of tension and with low viscosity of the muscle, i.e. the muscle can alter its length rapidly in response to alterations in tension. These alterations in length may be assumed to imply considerable metabolic activity and therefore considerable oxygen use.

The question of oxygen use was investigated by Evans (1923); according to him alterations in tonus (which I take to mean Winton's postural activity) do not in fact imply any great alteration in oxygen consumption. Out of five cases quoted in which the tonus was reduced, the oxygen consumption was lowered slightly in three, raised slightly in one and raised markedly (doubled) in one. Out of fourteen cases in which the tonus was increased the oxygen consumption was decreased slightly in eleven, much in one and increased slightly in two. Putting on one side the single experiment in which there were considerable changes both in increase and decrease, there remain seventeen cases in which the tonus was altered in one direction or the other and in which the alterations in oxygen consumption were trifling. Taking Evans's figures at their face value, the condition which he calls "normal", *i.e.* something between extreme tonic shortening and extreme elongation, was that in which the muscle used least oxygen. I imagine that he would not stress such a contention, much as it appeals to the imagination, for the simple reason that, as he points out, his tonus was not uncomplicated by rhythmic contractions, and these, because of their considerable demand for oxygen, rule out the possibility of making nice calculations as to the oxygen used in tonus.

It is necessary here to point out the difference between tonic elongation of the muscle and the sort of elongation which is produced by suspension of the metabolic process. Exposure of the muscle to an extremely low oxygen pressure or to very small doses of cyanide will cause the fibre to elongate, but such measures cut down the oxygen consumption.

Table XXVII

Exp.	Concentration of cyanide	Oxygen consumption c.c. per gm. per min.	
		Normal	After cyanide
20	$N/10,000$	0·31	0·22
21	$N/1860$	0·13	0·065
22	$N/930$	0·13	0·056
23	$N/10$	0·31	0·062

On the other hand, Evans states that in the tonic state, not more, but less oxygen is consumed, than by the same tissue when in the relaxed condition.

The following model (Fig. 102), which in essence is the same as the schematic one drawn by Winton, will illustrate the nature of the contraction, or at all events the relaxation of smooth muscle, and we may suppose that any nervous impulse which tends to shorten or lengthen the muscle must ultimately act upon one or other of the factors involved.

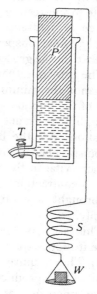

Fig. 102. Model to illustrate the properties of unstriped muscle. If a weight is placed in the pan at *W*, the pan will suddenly drop owing to the extension of the spring. If, further, the tap *T* is very slightly open, the weight will gradually descend further owing to the fall of the piston *P*. If the tap is opened fully so that the fluid gushes out, there will be a rapid descent corresponding to loss of peripheral tone.

Thus a relatively quick contraction would principally be tantamount to a sudden strengthening of the spring, a quick relaxation to weakening of the spring. This would probably be controlled by the motor nerve. A decrease of tone would correspond to a lowering of the specific gravity of the fluid in *B*, and increase of tone to the raising of the specific gravity—so that the piston floated to the top.

Whether the *nervus erigens* affects only the viscous or the non-viscous elements as well it is not possible at present to say. The simplest provisional hypothesis is that in the sub-maxillary vessels, to return to our original example, central tone involving the non-viscous phase is the concern of and can be increased by the sympathetic and peripheral tone, the viscous phase is an expression of the normal life of the cell and is modified by stimulation of the chorda.

(2) As regards the correlation of the function of the vessels with that of the gland cells, the position is as follows: maximal activity of the gland cells demands minimal contraction on the part of the vessels; the condition of peripheral tone is approximately appropriate to the resting metabolism of the gland, whilst the maximal degree of contraction of the vessels involves decrease of blood supply, which suspends the activity of the cells and is only to be justified on the ground of the blood being required by other organs.

The principle that the maximal degree of activity of the organ demands the minimal degree of contraction of the muscle in the wall of the arteriole is of course not confined to the salivary glands; it applies to tissues generally and its very universality raises the question: Are the blood vessels of all organs supplied with antagonistic nerves? If not, it would seem that the inhibitory nerve is more vital to the function of the organ than is the nerve which causes contraction, unless indeed inhibition of the vessel wall can be brought about by some agency other than that of its nervous supply.

This last hypothesis was rather fashionable at one time. Let me state it more explicitly: it was that, when organs became active:

(i) They required an increased blood supply.

(ii) They produced metabolic products which had a dilator action upon the vessels.

Therefore when the organ required blood it received it automatically. On the other hand, the reason for cutting the blood supply of the organ down to a minimum was that blood was required elsewhere, therefore "elsewhere" presented its demand to the brain, the vaso-constrictor nerves were stimulated and the blood supply in the resting organ was cut down.

By way of parenthesis let me point out that in this, as in so many other cases, the existence of vaso-dilator nerves and of metabolic dilator products are in no way necessarily alternative propositions.

If dilator nerves exist it does not disprove the possibility of a local chemical mechanism and *vice versa*.

The strong point about a local chemical regulation is its selfishness; if chemical dilator bodies are produced, it is not easy for the nervous system to deny the organ blood.

To sum up then, it would seem that the antagonistic nerves of the sub-maxillary vessels not only act essentially on different mechanisms in the cell, but form parts of different functional syndromes.

(3) The researches of recent years have shown that the antagonistic nerves are more widely distributed to the blood vessels than was at one time supposed. The skeletal muscles form about half the body, yet they were supposed only to be supplied with constrictor fibres, whilst the existence of vasomotor fibres to the heart and lungs was denied completely. I shall discuss the lungs elsewhere. Here I shall say something about the coronary system and about striped muscle.

The Coronary System

The organ which, in the highest interests of the body, could best justify complete selfishness is the heart. One might expect in the coronary system to find a local chemical dilator regulation most completely developed: in fact it would not be surprising if there were no other, and this was a view in which till recently many acquiesced. In the interests of "safety first", the occurrence of vaso-constrictor nerves might be positively dangerous, while vaso-dilator nerves might be superfluous. Let us see what the facts are.

I will pass very lightly over the early work, which led to the idea that the coronary vessels possess no nervous control. There seems to be no doubt now that up to that point research had simply failed. That adrenaline had marked effects was shown by Langendorff (1907) and everyone who succeeded him. The effect of this hormone was however in the direction of increasing and not of decreasing the blood flow through the vessels, and in those days the stimulation of sympathetic nerve endings was rather particularly associated in the minds of physiologists with pressor action.

The difficulty about the observations on the innervation of the coronary system has always been the elimination of the effects on it of the heart's own contraction. These effects are twofold. In the first place there is the actual mechanical effect on the coronary vessels of the rhythmic alterations in tension of the heart fibres. In the second place there is the possibility of metabolic products liberated by the contraction of the heart affecting the calibre of the coronary arterioles. Within recent years the whole field has been worked over very carefully by Anrep (1926) and his school.

As regards the direct effects of the heart's own contractions, each contraction of the heart squeezes the coronary vessels and therefore impedes the flow of blood into them. Given the same average pressure of blood at the entrance of the coronary arteries the blood does not flow through them in greater volume when the heart beats more rapidly. The general principle that metabolic products resulting from the heart's action tend to relax the walls of the coronary arteries is accepted by Anrep, both as regards CO_2 and other bodies, on the basis of work carried out by Dixon and myself (1907) for CO_2 and by Markwalder and Starling (1913). The acceptance of these facts only renders more urgent the necessity of their elimination in any satisfactory exploration of the nervous mechanism of the coronary system. Anrep has achieved this by the use of the innervated heart-lung preparation, in which the heart can be driven artificially at a constant speed, in which the pressure at the origin of the coronary arteries can be maintained constant, and in which the output is also invariable. Keeping the above factors constant the nerves to the heart may be stimulated, with the following results: In the heart as in the salivary glands the vessels normally show the phenomenon of peripheral tone, and hence the possibility of vaso-dilatation by the abolition of the degree of peripheral tone which prevails. This inhibition is brought about by stimulation of the sympathetic. So probably ends a long controversy which has largely centred about the effect of adrenaline on the coronary vessels. The literature on this subject may be found in the review by Anrep. The vaso-dilator action of adrenaline on the coronary vessels was first described by Langendorff (1907) in the fibrillating heart, and was further investigated by Brodie and Cullis in 1911. Anrep's pupils, Hammouda and Kinosita (1926), have made the physiology of the coronary system clearer by demonstrating this dilator effect as the result not only of reflexes from the sensory nerves, but of cerebral anaemia and asphyxia, influences which tend in the viscera to cause vaso-constriction and rise of blood pressure.

The degree of tone which usually prevails in the coronary vessels is, however, greater than that which would be present in the absence of all nervous stimulation; as in the case of the salivary glands, there is added to the independent tone of the vessels a certain degree of constriction caused by impulses reaching the heart via the vagus. These impulses can be much increased by active excitation of that nerve. Here again the most interesting facts about the vagus constriction of the coronary vessels have to do with the reflexes which affect it. Increase in the output of the heart will abolish the vagus tone of the vessels.

It is of interest too that the coronary system appears to have the property of "escape" from vagus constriction in the same sort of way that the heart itself has the power of "escape" from vagus standstill. Although the effect of the vagus on the muscle concerned is on the heart to produce inhibition and on the coronary vessels contraction, yet physiologically the vagus can no more bring the heart to a permanent standstill by the indirect method of strangulating its blood supply than by the direct method of permanent cardiac inhibition.

Vasomotor Nerves to Striated Muscle

Numerous experiments have been carried out in the intact animal in which a preparation has been made of the muscles of the calf of the leg, and in which the nerve going to them has been cut and stimulated. As regards the cutting of the nerve, it is agreed on all hands that this procedure produces a larger blood flow through the muscle and that, therefore, it abolishes a condition of central vascular tone. On stimulation, the results obtained have varied. Verzár (1912) for instance, in the cat, sometimes obtained a considerable decrease in the blood flow through the muscle when he tetanised it by stimulation of the nerve with a Faradic current. At other times he obtained a small increase, but in either case there was generally an increase after the tetanus passed off. The question of increase or decrease is, however, obscured somewhat by the very varying initial conditions of vascularity of the muscle.

As it was incredible that over any considerable time functioning muscle should have a smaller blood flow than resting muscle, the question was taken up again by Kato and myself (1915). We applied Faradic currents of the duration of 0·3 sec. every second for 15 consecutive minutes to the nerve with the result that the blood flow quickened. Here again the results were not uniform; though we never got a falling off of the blood flow during the period of activity, we only obtained a marked increase in three out of five experiments.

Anrep and his co-workers (1934) have recently taken the matter up on the same lines as those of their researches on the coronary circulation.

In the first place he confirmed the observation of Zuntz (1878) (which had only been partially confirmed by Barcroft and Kato) that section of the nerve is followed by an increased blood flow. Anrep however added the material fact that in the anaesthetised dog the increase is to be observed not only in the uncurarised, but also in the curarised muscle. The vessels of the gastrocnemius of the anaesthetised dog are then in a state of central tone which is removed when the nerve is cut. In the curarised muscle strong stimulation of the peripheral end of the sciatic causes constriction. This constriction is abolished by ergotoxine. Evidently the sciatic contains vaso-constrictor fibres to the gastrocnemius. So much then for the muscle at rest; now as to what can be learned about the vascular conditions during activity. Going back to the non-curarised gastrocnemius, if it be stimulated to give a short series of tetanic contractions, the blood flow is checked during the actual periods of contraction, and if the contraction be strong enough, there may be even an almost complete arrest. Between the contractions there is an increased blood flow. As has already been stated, the contractions may be abolished by curare and the vaso-constriction by ergotoxine. If then the nerve be stimulated, the result is dilatation.

These results form a very beautiful confirmation of Gaskell's position based on his observation that dilatation of the blood vessels of the mylohyoid in the curarised frog was to be seen under the microscope when the nerve going to it was stimulated. The facts with regard to Gaskell's experiment (1878) have never been challenged, but his interpretation of them was never accepted by Langley (1900) who, in his classical article on the sympathetic system in Schafer's *Text-book of Physiology*, took the view that there were not vaso-dilator nerves to muscle but that the dilatation was due to the formation of metabolites in the muscle. This explanation has not been excluded from Anrep's experiments, but his curves appear to have such sharp time relations as to throw the burden of proof on to him who would argue that stimulation of the nerve going to a curarised muscle produces increased breakdown in the muscle. In the blood vessels of the salivary glands, of the heart and of skeletal muscle we see the same picture. It consists of four principal figures, complete relaxation, peripheral tone, central tone, constriction.

In the last resort it is clear that the force which dilates the blood vessel is the pressure of the blood within it. This pressure is balanced against the tension in the wall. That tension can vary through the various stages from extreme constriction wrought by the constrictor nerve to considerable, perhaps complete, loss of peripheral tone wrought by the dilator nerve. As tone disappears, so the push of the blood produces a greater effect and the lumen of the vessel enlarges.

The Pupil

When we pass to the consideration of the calibre of the pupil we are confronted probably with a different proposition and certainly with a different mechanism.

The antagonistic mechanism of the iris presents an essential difference from that of the blood vessels. It is an antagonism not merely of two nerves, the third cranial nerve and the sympathetic, which act on the same muscle, but of two muscles as well, the circular sphincter which the third nerve supplies and the radial *dilator iridis* innervated by the sympathetic. The size of the pupil is the result of the algebraic sum of the tensions due to the antagonistic muscles. And this is the first mention that has been made of summation, or rather the summation of antagonistic influences. There is I believe no evidence of summation so far as the action of

the vagus and sympathetic on the pacemaker of the heart is concerned. We have seen nothing of the kind in our consideration of
the calibre of blood vessels; it would appear that the mechanism
by which two antagonistic nerves act on the same muscle does not
make for summation. Summation only appears when the antagonism of two muscles is concerned.

Granting that we have in the pupil a mechanism which we have
not met before we can scarcely fail to enquire its object. I suppose
that object to be "scope". The alternative would be for one or
other muscle to pull against an elastic system—something which is
fundamentally a spring: that alternative is adopted in the case of
the ciliary muscle which pulls against the tension in the choroid set
up by the intra-ocular pressure.

The very different degrees of illumination to which the retina is
subjected demand alterations in the diameter of the pupil which
are probably beyond the compass of a sphincter working against a
spring.

The Eye Muscles

The eye muscles bring us to a case of antagonism of a more complicated character than those already studied. Like the pupil the
system is not one in which two nerves act on the same muscle but
in which antagonistic muscles pull in opposite directions on one
structure and thus fix its position.

When the eye is fixed, the nervous processes would appear
superficially to be much the same as those which fix the pupil in a
certain degree of distension, *i.e.* a steady stream of impulses down
the nerve paths producing a steady condition of tone in the antagonistic muscles. When the eye moves, however, the increased
contraction of one muscle is associated with a reciprocal inhibition
in the antagonist. The eye ball therefore is fixed with the marvellous
accuracy which obtains, by an exquisite adjustment of the tensions
of the muscles which control it. These tensions are adjusted by a
system of antagonistic influences so poised, the one against the
other, as to maintain the fixity of the organ in a very exact position.
The object to be gained is not scope but delicacy and swiftness of
movement. Consideration of the eye muscles leads us up to yet a
new point, complementary rhythmic activity of the components.

To such a one as myself who enters the domain of the nervous system through the gate of respiration, the oculo-motor nucleus acquires quite a fresh interest when we see its fixity of poise broken down into a rhythm. Such is the case at the onset of nystagmus. The first point of interest about nystagmus is simply that it is a rhythm and the second that, remotely at all events, it is caused by the continuous play of sensory impulses upon the responsible mechanism in the central nervous system. No doubt the immediate sensations are intermittent and come from the eye muscles as part of the rhythm, but the remote sensation, that which actually breaks the tone into a rhythm, is constant. To take the most familiar case of lateral nystagmus, that of the man who looks out of the railway carriage window at the passing embankment—it is the sense of uniform movement of the object at which he is looking which causes the break of the tone into a rhythm. So too, I suppose, in other forms of nystagmus; when for instance one semi-circular canal is warmed, presumably the flow of impulses from it to the brain is steady but it breaks the fixity of position of the eyes into a rhythm. To say, however, that a condition of poise is broken into one of rhythm is a very imperfect description of what is taking place even on the surface in nystagmus. The poise was maintained by two antagonistic sets of nervous influences; each of these is broken into a rhythm, and moreover these rhythms alternate in a co-ordinated way so that, when the muscles pulling in one direction are contracting, their antagonists are relaxed and *vice versa*. Moreover, the two rhythms are not the same; one has a long contraction time and a short period of inhibition, the other a short contraction time and a long inhibition. Putting the relative lengths of time of the contraction and relaxation periods on one side, the important conception which arises is that of an *alternating* antagonism instead of a simultaneous one.

It must not be supposed that the mere movement of the eye to an extreme position to the left necessarily causes a sudden reversal to the right. It is possible to move the eye to the extreme position to the left and maintain it steadily in that position. It is apparently the impulse to keep the eye *constantly moving* to the left which involves this sudden reversal of movement.

Is nystagmus a rhythm which has become adapted or is it built up from some primitive process such as the animal watching itself go past the food which it wishes to strike, or watching the food go past it?

What is the relation, if any, between nystagmus and walking and respiration?

Joints

Consideration of the eye muscles has led me somewhat away from the main line of my treatment of antagonistic nerves; it has led to the very important subject of the establishment of rhythms. Let me now return.

There is nothing fundamentally different between the neuro-muscular mechanism of the eye and that of a joint. In one case the muscles are attached to a soft structure and in the other to a bone, but the principles are the same, muscles which pull against one another and are antagonistic mechanisms capable of contraction and inhibition, but the inhibition of the skeletal muscles is the mere cessation of contraction and is very different from that of an unstriped muscle fibre; it is that no impulses pass down the nerve and no motor units contract; it is a silence, and that silence is something induced in the central nervous system as the result of reflexes playing upon it; the inhibition of skeletal muscle has nothing in common with the removal of peripheral tone caused by stimulation of an inhibitory nerve.

Reviewing the whole field of antagonistic nerves, now that the ground has been surveyed it presents a very different picture from that which it suggested to Gaskell.

It involves no common principle, either in mechanism or in function, it is merely the working out of a mechanical device which is capable of great scope and refinement. At one time it is invoked to solve the paradox of opening the blood vessels at the moment of muscular contraction, at another to provide scope of movement in the pupil, and at a third to implement the exquisite adjustment of the finger to the thumb.

REFERENCES

ANREP, G. V. (1926). *Physiol. Rev.* **6**, 596.
ANREP, G. V., BLALOCK, A. and SAMAAN, A. (1934). *Proc. Roy. Soc.* B, **114**, 223; and personal communication.
BABKIN, B. P. and McLARREN, P. D. (1927). *Amer. J. Physiol.* **81**, 143.
BARCROFT, J. and DIXON, W. E. (1907). *J. Physiol.* **35**, 182.
BARCROFT, J. and KATO, T. (1915). *Phil. Trans. Roy. Soc.* B, **207**, 149.
BRODIE, T. G. and CULLIS, W. C. (1911). *J. Physiol.* **43**, 313.
DE BOER, S. and CARROLL, D. C. (1924). *Ibid.* **59**, 312.
EVANS, C. LOVATT (1923). *Ibid.* **58**, 22.
GASKELL, W. H. (1878). *Ibid.* **1**, 262.
—— (1900). Schafer's *Text-book of Physiology*, **2**, 220. Edinburgh and London.
GASSER, H. S. (1926). *J. Pharm. Exp. Therap.* **27**, 395.
GUNN, J. A. and UNDERHILL, S. W. F. (1914). *Q.J. Exp. Physiol.* **8**, 275.
HAMMOUDA, M. and KINOSITA, R. (1926). *J. Physiol.* **61**, 615.
LANGENDORFF, O. (1907). *Centrb. Physiol.* **21**, 551.
LANGLEY, J. N. (1900). Schafer's *Text-book of Physiology*, **2**, 640. Edinburgh and London.
MARKWALDER, J. and STARLING, E. H. (1913). *J. Physiol.* **47**, 275.
VERZÁR, F. (1912). *Ibid.* **44**, 243.
WINTON, F. R. (1930). *Ibid.* **69**, 393.
ZUNTZ, N. (1878). *Berl. klin. Woch.* **15**, 141.

CHAPTER XIII

THE PRINCIPLE OF MAXIMAL ACTIVITY

Sometimes as I have stood contemplating the majesty of a locomotive by the platform of a railway station, I have thought how meaningless would be the machine unless considered in respect of its activity, unless we think of it in terms of horse-power, or of the rate at which it will pull a certain load or in some such terms; apart from the fulfilment of its function, the engine is an agglomeration of curiously shaped pieces of metal. The condition of exercise is not a mere variant of the condition of rest, it is the essence of the machine.

To what extent do the same considerations apply to the body? In what terms are we to describe organs of the body, in terms of rest regarding exercise as a variant, or in terms of its capacity to do work?

If we revert to the engine and consider its "metabolism", the coal consumption, or oxygen consumption, when the locomotive is exerting its full power, is probably a pretty definite figure. Call that A. The consumption of fuel (B) when the engine is at rest is a quite indefinite affair and might vary from nothing to a value limited by the amount of steam which the safety valve could pass. In the case of the engine then, if B has no precise meaning, the ratio A/B will be in the same category and $A-B$ will only have a meaning if B is small as compared with A, i.e. if $A-B$ is approximately A.

In the present chapter I have gathered up some reflections which have passed through my mind at different times as to the justice of expressing the metabolic activity of an organ in terms of A, or of $A-B$, or of A/B, the last being the classical method and the one which I shall consider first.

My earliest physiological venture was an enquiry into the degree of metabolic activity which the sub-maxillary gland underwent when stimulated. This enquiry was followed by similar researches on other organs, some undertaken by myself, some by other workers in the laboratory. The results were summarised in a table now to be found in many text-books (Table XXVIII). It probably still stands as being fairly just and gives the impression that the metabolism of a good many organs during activity is about three times that during rest.

Table XXVIII

Organ	Condition of rest	Oxygen used per min. per gm. of organ c.c.	Condition of activity	Oxygen used per min. per gm. of organ c.c.
Voluntary muscle	Nerves cut. Tone absent	0·003	Tone existing in rest. Gentle contraction. Active contraction	0·006 0·020 0·080
Unstriped muscle	Resting	0·004	Contracting	0·007
Heart	Very slow and feeble contractions	0·007	Normal contractions. Very active	0·05 0·08
Sub-maxillary gland	Nerves cut	0·03	Chorda stimulation	0·10
Pancreas	Not secreting	0·03	Secretion after injection of secretin	0·10
Kidney	Scanty secretion	0·03	After injection of diuretic	0·10
Intestines	Not absorbing	0·02	Absorbing peptone	0·03
Liver	In fasting animal	0·01–0·02	In fed animal	0·03–0·05
Suprarenal gland	Normal	0·045	—	—

The statement which I have just made brings back mental reservations which perhaps haunted me too little at the time and which have grown rather than diminished as time has passed. Note the phrase "the metabolism...during activity is about three times that during rest". The resting metabolism is taken as the standard, and that at the height of activity is expressed as a multiple of the resting value. Yet the impression which I have carried away from these researches is that it is much more easy to obtain uniform values for active than for resting organs. The following series

(Table XXIX) will illustrate my point. They are data which were obtained by Brodie and myself on the effect of certain crystalline diuretics on the kidney (Barcroft, 1908).

Table XXIX. *Oxygen used per gram of kidney per minute*

No.	B Rest c.c.	A Diuresis c.c.	$A/B \times 100$	Diuretic
1	0·018	0·064	356	NaCl
2	0·016	0·081	506	Urea
3	0·075	0·095	122	Urea
4	0·015	0·075	500	Na_2SO_4

Evidently the great variation in percentage increase is due to the large variation in the resting figures. Would it not be fairer to say: "The kidney can be pushed up to a metabolism of 0·08 ± 0·016 c.c. of oxygen per gram per minute? Short of that it may be found in almost any state of activity down to 0·015 c.c. and possibly below". What the kidney is doing on the occasions when it is apparently resting but using very varying amounts of oxygen, I do not know. I will pass on to the reader a phrase which I once heard from the lips of van Slyke—"The metabolism of the kidney appears largely to be concerned with its own intracellular affairs". But I have never been able to free the "resting" kidney from these irregularities. The following figures (Table XXX) represented a very definite effort on the part of Mrs Tribe (now Mrs Oppenheimer) and myself (1916) to do so, but even here there is a threefold variation.

I am not suggesting that there is any real mystery. Doubtless the problem is just one of considerable complexity which would yield to sufficient skill and patience. In the meantime perhaps the most obvious line of enquiry would be that of asking whether the minimal of resting metabolism was regulated by the blood flow.

There is a rough correlation in the above table between the blood flow and the oxygen consumption, but clearly the correlation will not stand in detail. Hayman and Schmidt (1928), who worked on a much more extensive scale, seem to have had similar experience.

Table XXX. *Data of vascular and metabolic conditions of rabbit's normal kidney*

No.	Weight of rabbit kg.	Weight of kidney gm.	Urine flow, c.c. per min.	Blood-pressure mm. Hg	Blood flow through kidney, c.c. per gm. per min.	O_2 used by kidney, c.c. per gm. per min.	Hb value % of normal human	Saturation of venous blood %
1	—	6·8	0·2	105–98	1·7	0·057	67	68
2	—	4·3	0·05–0·01	128–120	0·9	0·074	88	39
3	—	6·5	Slow	154–127	0·9	0·046	70·3	51
4	2·6	10·0	Slow	105–98	2·0	0·063	118	59
5	1·5	5·0	0·05	100–108	3·7	0·164	—	—
6	1·6	7·7	0·02	104	2·6	0·100	—	—
7	3·6	15·0	0·02	106–94	2·0	0·072	97	77
				Mean	2·0	0·082		59

Their own words apply equally to the above table and to the results of experiments which they describe: "Nor did any single experiment show a uniform relation between increase of blood flow and increase of oxygen consumption. But when the evidence of all the control observations is considered a fair degree of correspondence is shown between blood flow and oxygen consumption".

On the other hand Bainbridge and Evans (1914) obtained quite a different result. The oxygen consumption remained relatively steady at an average of 0·046 c.c. per gm. of kidney per min., whilst the blood flow varied fourfold, viz. from 12·5 to 55 c.c. per min.

In comparing the experiments of Bainbridge and Evans with those of Hayman and Schmidt and of Mrs Tribe and myself, it is at once evident that the former were carried out in a preparation freed from control by the nervous system, whilst in ours the kidney was to some indeterminate extent under its domination. The immediate effect of cutting from nervous control such a preparation as that of Bainbridge and Evans would be to increase the blood flow through it. It might be argued with some degree of reason that our preparations were in a degree of partial asphyxiation which, indeed, may be the normal condition of the so-called resting kidney. I may draw the picture in a little more detail, though it is a hypothetical picture. The kidney in one of Hayman and Schmidt's experiments may be visualised as under nervous influence; seen in the laboratory in which the experiments were conducted, the nervous influence would appear as the constriction of the vessels to certain tubules (I include the glomerulus in this term) and therefore the oxygen consumption would be cut down to the extent consequential to the shutting off of blood. In Bainbridge and Evans's experiments the blood flow is in all cases adequate, and therefore an alteration within their limits is immaterial to the oxygen consumption.

But even accepting the Philadelphia picture it must not be forgotten that diffusion takes place within the kidney itself, and that the effect (on the oxygen consumption of the kidney) of cutting off the flow from every alternate tubule might be quite different from the effect of cutting it off from every alternate pyramid.

The reader may think that I am drifting into a dissertation upon the physiology of the kidney. That is not my intention, I am merely trying to illustrate the difficulty of defining any basal condition of kidney metabolism.

Is the basal condition that of Bainbridge and Evans's experiments, in which the blood supply is adequate and the metabolism is uninfluenced by its further rise, or is it basal when the kidney is being partially throttled, as is very likely normally to be the case? If the basal condition is one of partial throttling, what is the degree of that throttling? Is it such that further throttling would produce irreversible injury?

The endeavour to obtain basal figures for resting muscle has met with no more success. Perhaps the most definite, also technically the most perfect, effort in this direction is that of Himwich and Castle (1927) given in Table XXXI. The highest of their figures is more than 250 per cent. of the lowest. This is not to be accounted for by variations in the blood flow nor of the temperature of the muscle.

Even greater variations were found by Langley and Itagaki (1917), who researched on cat's muscle. The following is their comment: "The unsatisfactory feature of the experiments is the variation in the oxygen use found in some cases from successive samples of blood coming from the same muscles". The term "oxygen use" as defined by them means "the oxygen used by unit weight of substance in a unit time". It varied in their experiments from 0·0013 to 0·0076 c.c. per gm. per min.

A little earlier Toyojiro Kato and I (1915) had made an assault on the gastrocnemius of the dog, in which we also had hoped to get a basal metabolism for muscle with which to compare our data obtained from the active organ. We made the following statement:

Having obtained the qualitative result that the coefficient of oxidation rises with activity, it is difficult to pass from that to any quantitative statement as to how much the oxygen consumption of the muscle increases for a given degree of activity. At present we can only point out some of the difficulties which must be overcome before any such calculation can be made. Perhaps the most fundamental of these centres about the coefficient of oxidation of resting muscles.

Table XXXI. Simultaneous oxygen consumption of isolated muscles and of entire animal

Exp.	Weight of animal kg.	Temperature of muscle °C.	Weight of muscle gm.	Blood flow per hour per gm. c.c.	O₂ utilised from 100 c.c. blood c.c.	O₂ consumption per hour per gm. muscle c.c.	O₂ consumption per hour per gm. entire animal c.c.
17	23·31	39·4	110	4·08	5·33	0·217	0·508
	23·31	40·3	110	3·72	6·19	0·230	0·568
18	16·12	38·0	72	9·42	2·44	0·230	0·447
	16·12	38·0	72	6·42	3·75	0·241	0·461
20	14·57	41·7	64	7·98	7·22	0·576	0·678
	14·57	37·0	64	6·84	7·96	0·544	0·636
21	29·70	37·5	90	5·34	4·43	0·236	0·404
	29·70	38·3	90	4·62	4·79	0·221	0·495
22	26·80	39·6	112	5·34	8·62	0·460	0·524
	26·80	37·5	112	4·74	9·95	0·472	0·510
	26·80	37·5	112	4·02	11·13	0·448	0·506
	26·80	37·8	112	4·20	9·55	0·401	0·496
				Averages		0·356	0·512

In order to estimate the effect of functional activity upon oxidation, two obvious courses present themselves:

(1) To measure the total oxidation actually taking place in the muscle.

(2) To measure excess of oxidation during the stimulation and post-stimulation periods, over that which occurs in the resting muscle.

Whichever of these methods we tried, we found that we had no certain base-line from which to work. It may here be well to anticipate what we are going to say later so far as to state two facts:

(i) That in common with all other observers we have found that the coefficient of oxidation in resting muscle is very different in different muscles; and

(ii) That the increased oxidation caused by fifteen minutes' stimulation such as we have given may last for several hours, perhaps four or five.

Taking these facts together it seems quite probable that the oxidation which observers have been accustomed to regard as that of resting muscle is largely the after-effect of previous contractions, and in order to make any just estimate of whether the muscle was a fit subject for estimation of the kind, one would have to know its history for some hours before the experiment took place.

To turn now to the facts which have led to the above conclusions. Chauveau and Kauffman (1887) obtained figures varying from 0·0082 to 0·0029 c.c. per gm. per min. in three experiments; Verzár (1912), in eleven experiments, gives figures from 0·0086 to 0·0023.

In five experiments on muscle not artificially stimulated we obtained data which varied from 0·055 to 0·0052. These we append (Table XXXII), together with some remarks upon the extent to which the muscle could really be regarded as resting, to judge from its present and previous history.

It will be clear that those muscles in which the nerve was cut some time previously had a lower coefficient of oxidation than those in which it was intact. We must qualify this statement by noting: (1) that the muscles were not the same—in the one case the gastrocnemius was used, in the other the anterior belly of the digastric—it may be that these two muscles have different co-efficients of oxidation in any case; (2) that Zuntz (1878) describes mere cutting of the nerve as reducing the coefficient of oxidation by removal of the tone from the muscle. The experiments of Zuntz on the subject were few, and not altogether concordant. They were never completed, and it is very desirable that in the light of more recent work the matter should be taken up again.*

Table XXXII

Exp.	Muscle	Oxygen used per gm. per min. c.c.	Remarks
8	Digastric	0·030	The muscle had been stimulated a short time before the sample was taken
9	Digastric	0·055	The muscle from time to time underwent rhythmic movements
10	Gastrocnemius	0·017	Just before cutting nerve
11	Gastrocnemius	0·0052	Nerve cut 1 h. 40 m. before observation
12	Gastrocnemius	0·010	Nerve cut 1 h. 5 m. before observation

In the above discussion I suggested as a possible reason for the discrepancies found in the "metabolism of resting muscle", that the post-contractile oxidation might persist for a considerable time in our preparations. Another possibility would now have to be considered, namely that some degree of post-operative irritation of the end of the cut nerve maintained occasional contraction of the fibres. Were the work repeated it would be desirable to carry it out

* As the persons concerned have all been dead for many years I may be excused for recording the following as illustrating the degree of discipline which existed in some laboratories in the nineteenth century.

Being struck with the fact that so important a research had been left by Zuntz in an obviously fragmentary condition, I asked him one night when we were in Teneriffe why he had never finished it. His reply, as nearly as I can remember it, was as follows: "When I commenced those experiments, I was assistant to Pflüger at Bonn. Pflüger came round one day and finding me at work said, 'What are you doing?' I replied, 'I am testing the effect of abolition of tone on muscular metabolism'. Pflüger said, 'Well, but you have not asked my permission to do this. Either you must stop these experiments or leave my laboratory'. I was not in a position to leave the laboratory, so I stopped the experiments".

in association with Adrian's technique in order to secure that the muscle really was at rest.

A similar story may be told about the sub-maxillary glands (Barcroft and Kato, 1915).

The resting values in four experiments are:

Exp.	2	3	4	5	6	7
Coefficient of oxidation c.c. per gm. per min.	0·023	0·026	0·027	0·01	0·020	0·024

These apart from the inevitable exception—Exp. 5—are very uniform, but the inevitable exception cannot be ignored. With these examples for the resting gland (both the chorda tympani and the vago-sympathetic had been cut) we may compare the effect on the metabolism of what is probably a maximal stimulus—the injection of 0·5 mg. of pilocarpine per kg. weight of dog. This was done in Exps. 3 to 6; the following table shows in addition to the resting metabolism, the average metabolism for an hour after the pilocarpine was injected.

Exp.	... ...	3	4	5	6
Coefficient of oxidation c.c. per gm. per min.	{Resting B	0·026	0·027	0·010	0·024
	{Secreting A	0·048	0·55	0·052	0·65
	Ratio $A/B \times 100$	187	204	520	271

Here again (A) the metabolism during high activity is much the more constant figure. It is however fair to emphasise that the values for the active gland are integrated from a graphic representation of the data, whilst the resting data are those taken at certain moments. We do not know what the result would be of taking estimations at ten-minute intervals for an hour from the so-called resting gland.

It seems that the most uniform measure of an organ is in terms of the maximum which it can accomplish. And why not? The purpose of the organ is to do work and not merely to vegetate.

The matter may be considered from another angle. A fact which surprised many persons thirty years ago was that many organs appeared to be quite unnecessarily large.

The kidney may be taken as an example (Bradford, 1891, 1899). Bradford showed that two-thirds of the whole kidney substance may be cut away without any apparent ill effects. "The excision of approximately two-thirds of the total kidney weight is followed by a considerable and practically permanent increase in the amount of urine passed, but there is no considerable permanent increase in the amount of urea."

When these experiments were published they naturally attracted great attention, and wonder was frequently expressed that the kidneys were so much larger than was necessary. Since that time it has been shown by Ambard and Papin (1909) for urea and total sodium, and Davies, Haldane and Kellaway (1920) for bicarbonate —that the kidney can only secrete urine of a certain limiting concentration, which is independent of the rate of excretion of urine. Presumably this is true of each kidney (it may even be true for each tubule) just as there is for each arm a limit to the weight that its muscles can lift. The point therefore in having so great an amount of kidney would centre about the length of time necessary to free the blood from an abnormal content of some substance, when the kidney was secreting it at the maximal concentration.

I would like at this point to consider the many organs which may be partially removed without serious interference, and after learning how to take the measure of these organs when functioning at their maximum, to give some statement as to the relation between the normal degree of function and the maximal degree and equate that against the proportion of the organ which the body found essential. The data for so doing are however not at hand. Such would be the suprarenals, the pituitary, the islet tissue and the like.

The lungs may be considered in some detail. Before the end of the nineteenth century several workers had studied the effect of excision of one lung, in animals, or of its absence in man and with varying results. Some account of the earlier work will be found in the paper of Vaughan Harley (1899), who, in acute experiments on dogs, found an increase both in the total ventilation and the total metabolism. The matter was later taken up by Carpenter and Benedict (1909) who, working on a man with practically complete destruction of one lung, found that the metabolism was neither markedly greater nor markedly less than that of other persons of similar physique. In short, for sedentary purposes such as ruled during the investigation in question, it is immaterial whether the subject has one lung or two.

I do not know what the minimal amount of lung consistent with life may be in man, but I have seen goats to all appearance perfectly well when three-quarters of their pulmonary tissue was solidified—as shown by post-mortem examination.

Let me pursue the matter a little further. By the injection of starch or oil into the jugular vein, large quantities of the pulmonary vascular system may be completely occluded, as was shown by Dunn (1919). Yet after a derangement lasting only a short time, there is no apparent effect upon the circulation. As much blood as before traverses the lungs and it does so without any rise in the pressure on the right side of the heart. The only conclusion possible is that, previous to the injection, considerable vascular areas were denied, partially or completely, to the blood stream, and that after the injection these or some of them opened out to form channels for the blood which was dammed back by the starch or oil emboli.

The lungs are much too large for the animal as it stands. Nor is it a sufficient explanation that a margin is necessary for illness. The fact of course is that the lungs are not built for the animal "as it stands" or lies, they are built to subserve the needs of the animal when it is putting forth the utmost exertion of which it is capable. It is not unlikely that the resistance of the vascular bed of the lung, when opened out to the full, is one of the factors which limit the amount of work of which the animal is capable. Clearly if an endeavour is made by the heart to drive more than a certain amount of blood through the lung the pressure in the right side of the heart must rise. Does such a rise lead to distress, of which the probable expression would be breathlessness? That is a question which we may leave over till we have considered the heart.

In recent years opinion with regard to the lung has drifted rather in the direction of judging that organ in terms of its absolute capabilities—a rough and ready measurement of which is obtained from the vital capacity. This test has proved to be of great service in the selection of pilots for the British Air Force and I suppose for those of other countries. The point is a little interesting because the ultimate object of the test is to pick out the men who will have most command of their mental faculties when flying—which is not necessarily the same thing as to pick out the men who can run fastest or perform some other physical feat that at first sight appears to be more nearly related to the capacity of the lungs. Other things being equal the man with the biggest vital capacity for his size is likely to be the best pilot.

The vital capacity has become important in another connection. Let me quote from Macleod (1930):

Recently Peabody and Wentworth have called attention to the fact that the tendency of patients with heart disease to become dyspneic more readily than do healthy subjects depends largely on their inability to increase the depth of the respiration in a normal manner. This inability to breathe deeply corresponds to a diminished vital capacity of the lungs, that is, the volume of the greatest possible expiration after the deepest inspiration. In normal adults the following averages (Table XXXIII) were secured from a large series of clinical cases. The subjects are grouped into two classes, each group being subdivided according to height.

It would appear that in normal people the vital capacity is at least 85 per cent., and almost always 90 per cent. or more, of the standard adopted for each group. In elderly persons a slight decrease from these standards may be expected. Table XXXIV shows that there is a remarkably close relationship between the clinical condition of cardiac patients, particularly as regards the tendency to dyspnea, and the vital capacity of the lungs. Peabody and Wentworth believe that the determination of the vital capacity affords a clinical test as to the functional condition of the heart, since compensated patients who do not complain of dyspnea on exertion have a normal vital capacity. Patients with more serious disease in whom dyspnea is a prominent symptom, have a low vital capacity; and the decrease in vital capacity runs parallel with the clinical condition. As a patient improves, his vital capacity tends to rise; as he becomes worse, it tends to fall. In other diseases in which mechanical conditions interfere with the movements of the lungs, the tendency to dyspnea corresponds closely to the decrease in the vital capacity. The cause of the decrease in the vital capacity of the lung in cardiac decompensation is difficult to explain satisfactorily. It may be the limitation in the movements of the lungs produced by engorgement of the pulmonary vessels, by the weakness of the intercostal muscles, the rigidity of the bony thorax, emphysema, or accumulation of fluid in the pleural cavities. In cardiac disease the air in the lungs at the end of a normal expiration is usually increased. This is similar to the condition which attends exercise, and is probably a physiological adaptation to give optimum aeration to the blood, as explained above.

It has become more and more evident, since Peabody and Wentworth's researches, that a determination of the vital capacity is of great importance in the diagnosis and prognosis of several diseases, including heart disease and tuberculosis. It is also important in gauging the effects of treatment.

Table XXXIII. *The vital capacity of the lungs of normal males*

Group	No. studied	Height in feet and inches	Normal vital capacity c.c.	Number within 10% of normal	Highest vital capacity	Lowest vital capacity	Highest %	Lowest %	Number below 90% of normal
I	14	6' +	5100	9	7180	5030	141	99	0
II	44	Over 5' 8½" to 6'	4800	41	5800	4300	121	90	0
III	38	5' 3" to 5' 8½"	4000	31	5080	3450	127	86	1

The vital capacity of the lungs of normal females

Group	No. studied	Height in feet and inches	Normal vital capacity c.c.	Number within 10% of normal	Highest vital capacity	Lowest vital capacity	Highest %	Lowest %	Number below 90% of normal
I	10	Over 5' 6"	3275	5	4075	2800	124	86	2
II	13	Over 5' 4" to 5' 6"	3050	9	3425	2660	112	88	2
III	21	5' 4" or less	2825	16	3820	2500	135	89	1

In order that the value actually found in a given patient may be compared with the value which a healthy individual of the same body build would give, it is necessary for clinical purposes that some reliable and yet simple method be available from which the normal value may be computed. Lundsgaard and van Slyke and Dreyer have worked out several ratios for this purpose, and West has shown, by observations on 129 persons, the most useful of these is one based on the body surface. The body surface is determined from measurement of height and weight by the graphic chart of Du Bois....

Table XXXIV. *The relation of the vital capacity of the lungs to the clinical condition in patients with heart disease**

Group	Vital capacity %	No. of cases	Mortality %	Symptoms of decompensation %	Working %	Remarks
I	90–	25	0	0	92	Few symptoms referable to heart
II	70–90	41	5	2	54	History of dyspnea with exertion, yet able to do moderate work
III	40–70	67	17	89	7	Dyspnea with moderate exercise. Few able to work
IV	Under 40	23	61	100	0	Bedridden, with marked signs of cardiac insufficiency

* Certain cases were tested several times and, owing to changes in the vital capacity they appear in more than one group. In the "mortality" column they are included only in the lowest group into which they fall, "Symptoms of decompensation" indicate dyspnea while at rest in bed or on very slight exertion. Under "Working" are included only those actually at work, and able to continue. Many other patients in Group II were able to work but they are not included as they were still in the hospital.

The average value for vital capacity (in litres) divided by the body surface (in square metres) is 2·61 per sq. m. and for women 2·07.... The deviation from the values should not be beyond 15 per cent., the great proportion of normal individuals being within 10 per cent. of the above averages. Athletes give decidedly higher values and old people give lower ones.

THE HEART

The demonstration that the heart is larger and more powerful than suffices for the needs of the resting animal is of a different nature from the corresponding demonstration for the lung. Clearly it is not possible to reduce the heart piece by piece till no more is left than is required for the ruling circulation. On the other hand the demonstration is simple in another way, because the function of the organ is to do mechanical work and the work that it does is a quantity numerically calculable. Whether you consider the actual work accomplished or the energy put forth in doing it, it is clear that when the body is at rest the heart is only realising a fraction of its potentialities. Certainly as regards the heart and lungs the tendency in recent years has all been in the direction of the assessment of the cardiovascular system in terms of its capabilities. All modern research has gone to show the fallibility of judging the heart entirely from the sounds which it emits. I do not know quite when the change took place but I associate it in my mind with the names of Sir James Mackenzie and Sir Thomas Lewis. An appraisement of the heart in terms of what it can do is not without its difficulties, of which the first perhaps is to find suitable units in which to express its activity. Naturally the pulse comes first under consideration because of the ease with which it may be counted. But the function of the heart is to drive the blood round the body. The actual work performed per minute by the organ depends almost entirely upon the minute volume of blood which it expels and the pressure against which that blood is driven from the heart. What information does the pulse give about either of these quantities. Something no doubt; but I think it would be a bold as well as a skilful physician who would be prepared to turn such knowledge into units and correctly to calculate the work of the heart from his estimate of the arterial pressure and the stroke volume as appreciated by feeling the pulse.

A computation of the relative amount of work performed by the heart would be easier could we accept the gospel of the constancy of the stroke volume. This is not the place to argue the pros and cons of that gospel, but I make no doubt that while the stroke volume may be fairly constant over a certain range of medium output, it becomes very inconstant when the upper and lower limits of the heart's output are approached. The minute volume of blood if it is small (or in Krogh's phrase if it is below the point of adequate filling) is probably almost independent of the pulse rate. What happens to the minute volume in the most rapid heart beats is a point

on which little information is available. In violent exercise it is
not possible to make the necessary determinations; experiments on
paroxysmal tachycardia point to the possibility of the heart being
able to beat so fast as to become quite inefficient.

If we want to form some estimate of the heart from what it can
do we must look in another direction. It is not the smallest of
Lewis's triumphs that he and his school have elaborated the standard
heart test* on the basis of pulse counting. But, between the test,
and any numerical statement of blood driven round the body there
is a wide and imperfectly charted gulf. Correct judgment of the
heart's potentialities came into great prominence during the War
because it was necessary to form an accurate judgment on such
matters as whether recruits or "disabled" soldiers were or were not
fit for active duty. In this connection much was heard in England
of a condition known as Disordered Action of the Heart—known
for short as "D.A.H." This was a state in which a trivial amount
of exercise caused an undue rise in the pulse, which rise was sustained for an abnormally long time. The degree of breathlessness
was also out of proportion to the exercise taken. Here is something
more illuminating. Can the cardiac and respiratory systems be considered as parts of a single functional unit? They appear to be
capable of consideration in that light, and the term "effort syndrome"
was introduced by Lewis to signify the synchronous and associated
changes which take place in the cardiac and respiratory systems
during exercise. The effort syndrome may be judged from the
measurement of minute volume not of blood traversing the heart
(about which in D.A.H. cases we know almost nothing directly) but
of air traversing the respiratory passages. Incidentally it is breathlessness which in the eyes of the patient limits his capacity for
exertion.

Our own approach to the subject (I say "our own" because
Major Hunt and Miss Dufton† were members of my staff in
Chemical Warfare Organisation) was definitely not from interest
in the heart but from interest in the lungs, and the position seems

* I allude of course to the return of the pulse to approximately the resting
value in half a minute.
† Now Mrs Wilson.

to be that breathlessness may be due to either cardiac deficiency or pulmonary deficiency, in either case it is the result of a reflex and therefore it involves the condition of the brain. The third possibility exists namely that of purely central abnormality causing the usual afferent impulses to produce an abnormal efferent response.

If there is a cardiac deficiency so that the heart sends too little blood to the brain, the irritability of the brain will increase and the normal afferent impulses will produce both a too rapid heart and a degree of breathlessness. Similarly if there is pulmonary deficiency and the brain is provided with improperly aerated blood the reflexes will be heightened and the same results will follow, as Haldane pointed out. Up to this point the cardiac and pulmonary mechanism necessarily go hand in hand—but of course there are other possibilities, the mischief may be purely peripheral. If, say, the nerve endings in the heart stretch to an abnormal extent, why should the respiration be affected, or if an inspiration stimulates an unusual number of fibres, why should the heart be affected? The answers depend upon irradiation at the centres; that the respiratory and cardiac centres are harnessed together is pretty clear.

But to return to Major Hunt and Miss Dufton's (1919) experiments. They tested the disability of soldiers in terms of the degree of breathlessness on exercise.

It should be understood that these experiments made no profession of supplying an easy routine test, for they required great care and involved the training of the subjects to the point of being able to breathe into Douglas bags without upset to their respirations. The patient was placed on a bicycle ergometer, he breathed into a Douglas bag for four minutes before taking exercise, he then carried out a standard exercise for one and a half minutes, during which time his expired air was collected; it was also collected into separate bags during each successive minute of the first five minutes after the exercise.

It is perhaps more difficult to know how to treat the data so obtained than to obtain them. One very simple way is to neglect the ventilation for the first minute after exercise and, taking the subsequent minutes, i.e. the 2nd, 3rd, 4th and 5th, to compare the total ventilation of the man over the sum of these with that over the four

minutes before the exercise. Incidentally we have departed from
our ideal of estimating simply the maximal effort of which an
organ is capable and are now considering not an absolute measurement but a ratio. To this point I shall return.

In theory the next step would have been to give each person as
much exercise as he could take before he got completely out of
breath and to note the amount of exercise necessary. In practice
such a procedure was impossible because it might have injured the
unfit—a less draconic test was required and was easily worked out:
the men, trained soldiers, healthy untrained civilians, cases suffering
from D.A.H., from gas poisoning and from other disabilities were
exercised up to the point at which they could be grouped into
categories. The group into which any man fell usually was pretty
clear from an observation of the exercise necessary to double his
total ventilation over the time intervals used. The exercises given
were:

Table XXXV

Work done in 1½ min.	Exercise					
	A	B	C	D	E	F
Foot-pounds	2900	3800	5600	7020	8740	10330
Kg.-metres (approximate)	400	520	770	970	1200	1830

Now to illustrate what took place. Let us compare the healthy
trained and untrained men (Table XXXVI).

On the lesser exercises, A–C, it would be quite impossible to make
any judgment of the capacities of the two groups. At exercise D the
effect of training commenced definitely to show itself, but it was
not until F was reached that a clear line of demarcation came out.
The exercise does not double the ventilation of any trained man,
whilst amongst the untrained men, 114 per cent. is the smallest
increase in the respiration.

The mildest D.A.H. cases fell into a group in which exercise C
sufficed to double the ventilation whilst more serious cases of
dyspnoea, e.g. a nephritis case, showed a similar increment with
exercise B, in which the work done was about a third of the critical
exercise for the untrained men.

Now let me take up a point to which I said I would return. We departed from the ideal of considering the effect of exercise in terms of an absolute measurement and fell back upon a ratio.

We might have made comparisons of other characters, for instance we might have subtracted the resting value from the value during exercise and so found a figure for the effect of exercise, or we might have made no comparison at all, and merely taken the absolute figure during exercise. Taking no account of that during rest, the defence which we would have put forward at the time was

Table XXXVI. *Percentage increase in pulmonary ventilation after exercise*

Type	Exercise					
	A	B	C	D	E	F
Trained:						
F. W. R.	9	17	27	37	58	63
E. M. T.	16	28	33	46	59	77
J. M. S.	11	−12	33	32	46	54
G. O. T.	7	−9	20	35	36	55
J. W. H.	−3	6	15	31	42	74
J. R.	1	8	24	33	69	78
R. A. H.	7	10	19	31	57	71
	9	5	39	40	67	98
Untrained:						
S. W. G.	19	21	39	67	91	114
A. S. H. W.	12	20	39	52	70	121
J. H. S.	−2	10	35	47	78	148
F. D. H.	6	21	28	47	71	115

merely based on the nature of the data obtained. The ratio of the post-exercise value for the total respiration to the resting value is surprisingly constant, whereas each of the values themselves vary within wide limits. An example of this is seen in Table XXXVII. Does this adoption of the ratio mean that we have pulled down our flag? Have we ceased to consider the organ simply in terms of what it can do? I think not. The introduction of the ratio appears to me to be consequential on something else. We are discussing not simply what the organ is doing, but the effect of that upon what something else, namely the cardio-respiratory central mechanism, is doing.

We are considering something of the second order: it is that which introduces the ratio.

Table XXXVII

B Total ventilation in litres in 4 min. before exercise	A Total ventilation in litres in 2nd–5th min. after exercise	Ratio A/B
4× 6·87	30·55	1·11
9·12	37·10	1·02
9·56	39·55	1·03
10·23	37·10	1·02
10·23	44·00	1·07
Variation as a percentage of the lowest 49 %	44 %	9 %

That this is so may be shown by another typical example in which the condition of the respiratory centre is purposely altered by the administration of CO_2 and the effect of exercise is built upon that alteration.

Table XXXVIII

B Total ventilation in litres in 4 min. before exercise	A Total ventilation in litres in 2nd–5th min. after exercise	Ratio A/B
4× 6·51	37·35	1·37
6·70	36·35	1·34
7·81	49·10	1·57
9·16	52·10	1·42
10·85	64·95	1·50
11·12	67·89	1·52
Variation as a percentage of the lowest 86 %	78 %	15 %

It would seem that though the respiration is in different conditions under different circumstances, the effect of a certain exercise in the subject studied is to multiply the resting ventilation by $1·45 \pm 0·13$.

Now we come back to the return of the pulse to the normal as a method of testing the heart.

The test involves all the compromises which we have made:

(1) As a practical test, the substitution of reaction to a limited exercise, for the reaction to a maximal exercise.

(2) The substitution of observations made on the cardio-respiratory reflex apparatus for observations made on the heart itself.

(3) The use of the resting value as a base-line.

The justification of the first compromise is the opportunist one, that to push exercise to the limit might injure the patient. The justification of the second is: "what the heart does" is better tested by its effect than by evidence of a lesion, for on the one hand the heart may be incapable without there being any definite lesion and on the other a lesion may be compensated for to such an extent as to be compatible with capability.

The justification of the third compromise is that it is consequential on the second. There is, in this test, a fourth compromise, which from the practical point of view is the great merit of the test and the index of its ingenuity—in considering the *time taken to return* to the resting value, the actual divergence from the resting value is banished from the picture.

The effect of exercise upon the heart does not come in. Does a given exercise increase the pulse in a given ratio? or does the exercise augment it by a given number of beats? or raise it to a given absolute value? These matters are outside the actual test. The fact may be seen from Fig. 103 taken from Meakins and Gunson's (1917) paper.

Taking the normal pulses first, it is not possible from the data given to predict what the effect of exercise is, further than that it approximately passes off (*i.e.* within six beats) in the second half minute after the exercise ceases. In the case of the pathological cases quite clearly the return is less rapid.

Fig. 104 shows the effect of disease in a beautiful way because it gives data for the same hearts before and after the infection. The difference is quite clear cut.

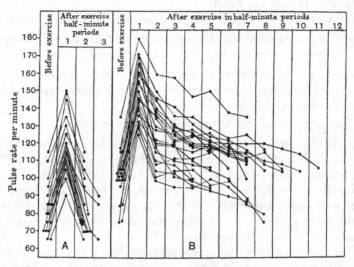

Fig. 103. A, curves showing normal return of heart rate to pre-exercise value; B, cases showing much slower return.

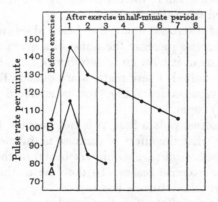

Fig. 104. Compound curves of four individuals, A, before, B, after an acute infection.

Admirable as is "the return of the pulse" test in practice, it would be intellectually more satisfying if we knew more about its real relation to the capacity of the heart for work. Yet admitting all the compromises which must be made and the possible gaps in the argument and crossing out the irrelevancies, the test, in so far as it is a test of the heart, yields a judgment of the organ in terms of the most it can do. Whether it is a test of the heart or not, it is a test of the physical capacity of the organism, and one the ultimate object of which is to inform the physician of the maximum effort which the organism can sustain.

The critical reader, at this stage, may ask me whether I have not reached an "impasse". He may say, "You are considering the body as made up of organs, say M, N, O, P and Q. You say that the development of Q depends upon the greatest strain that M, N, O and P can put upon it, but likewise the development of P depends upon the greatest strain M, N, O and Q can put upon it, and so with the rest. That is just going round in a circle. With such reasoning, why should the body not increase indefinitely?" Of course I admit the impeachment. The development of the body as a whole must surely be limited in some way by its environment or rather its ability to compete with its environment. Different forms of animal life have sought out different environments, and, having found them, depend one upon one faculty, another upon another faculty for his safety. The reader will easily guess what seems to me most likely about man. Man lives primarily by his intellect; his intellectual development to my thinking is conditioned by his capacity for the exact regulation of the properties of his "milieu intérieur".

To consider the constancy of man's temperature; if he were too small, his temperature would be too readily influenced by alterations of heat and cold in his environment; if he were too large he would be the victim of his own powers of heat production. There must be some critical size at which the two are most efficiently balanced. The same sort of considerations doubtless apply to other qualities, physical and chemical, of the blood and some compromise between these various values, critical for various properties, will be a potent factor in conditioning the order of size of mankind.

REFERENCES

AMBARD, L. and PAPIN, P. (1909). *Arch. Int. Physiol.* 8, 437.

BAINBRIDGE, F. A. and EVANS, C. LOVATT (1914). *J. Physiol.* 48, 283.

BARCROFT, J. (1908). *Ergeb. Physiol.* 7, 746.

BARCROFT, J. and KATO, T. (1915). *Phil. Trans. Roy. Soc.* B, 207, 149.

BRADFORD, J. R. (1891). *J. Physiol.* 12, xviii.

—— (1899). *Ibid.* 23, 415.

CARPENTER, T. M. and BENEDICT, F. G. (1909). *Amer. J. Physiol.* 23, 412.

CHAUVEAU, A. and KAUFFMAN, M. (1887). *C.R. Acad. Sci.* 104, 1126, 1763; 105, 328.

DAVIES, H. W., HALDANE, J. B. S. and KELLAWAY, E. L. (1920). *J. Physiol.* 54, 32.

DUNN, J. S. (1919). *Q.J. Med.* 13, 46, 129.

HARLEY, V. (1899). *J. Physiol.* 25, 33.

HAYMAN, Jr, J. M. and SCHMIDT, C. F. (1928). *Amer. J. Physiol.* 83, 504.

HIMWICH, H. B. and CASTLE, W. B. (1927). *Ibid.* 83, 104.

HUNT, G. H. and DUFTON, D. (1919). *Q.J. Med.* 13, 165.

LANGLEY, J. N. and ITAGAKI, M. (1917). *J. Physiol.* 51, 202.

MACLEOD, J. J. R. (1930). *Physiology and Biochemistry in Modern Medicine*, p. 647. 6th ed. London.

MEAKINS, J. C. and GUNSON, E. B. (1917). *Heart*, 6, 285.

TRIBE, E. M. and BARCROFT, J. (1916). *J. Physiol.* 50, x.

VERZÁR, F. (1912). *Ibid.* 44, 243.

ZUNTZ, N. (1878). *Berl. klin. Woch.* 15, 141.

CHAPTER XIV

DUPLICATION OF MECHANISM

Thirty years ago the meetings of learned societies were smaller than to-day, and their programmes less congested. The character of such meetings has undergone a consequential change. It was then the privilege, and to some extent the diversion of the, then, younger generation to listen whilst their elders engaged in controversy at considerable length and sometimes with no little emphasis. Such controversies often took the form of a discussion as to whether nature achieved a particular end by this means or by that. Looking back, perhaps the most remarkable feature of these battles was the assumption which underlay so many of them, that nature could have only one way of doing a particular thing, for on reflection it often turned out that the rival methods for the attainment of a particular object were in no way mutually exclusive. Thus if the kidney be proved to secrete it by no means follows that it cannot reabsorb. I remember putting this very possibility once to Langley in one of the chats which we used to have from time to time about secreting glands—I mean the possibility that the kidney could both secrete and reabsorb; to his generation such an idea appeared "woolly" and he at once said—"that is much too complicated": to which I could only reply that after all it seemed no more complicated than the kidney itself. It is on the whole a simple assumption that each histologically distinctive portion of the tubule has a different function from its neighbour. If you make that assumption, the possibility of secretion here and reabsorption there does not seem to strain the possibilities of the situation.

From the general *milieu* of discussion to which I listened the principal precipitate which settled out was I think the conviction that nature had more than one way of doing a great many things: and as a first approximation I think it is so.

One such discussion, I remember, occurred shortly after the

classical demonstration of the effect of CO_2 on the respiratory centre by Haldane and Priestley (1905). The point at issue was— What is the stimulus to the respiratory centre? The one view was that CO_2 was the stimulus. The protagonists of that view held that the centre was incapable, normally, of alterations in irritability. The opposite view was that the only stimulus to the respiratory centre was the apparent phase of the Hering-Breuer reflex and that the effect of CO_2 was merely to alter the irritability of the centre. Superficially it appeared more interesting that the mechanisms might exist side by side and each be the complement of the other. If the object in view was the crude one of increasing the total ventilation—say on exercise—there were two avenues of approach, either the nervous or the chemical. Each of these in the absence of the other would presumably be able to carry out the necessary regulation with some degree of efficiency. The body is therefore insured against the default of either mechanism.

But there are some finer points. Are the chemical and the nervous stimuli to the respiratory centre really, in the same sense, stimuli? or is one (the chemical) an increased lubrication of the machine, whilst the other (the nervous) is an increased effort on the part of the man who operates it? Of what is each stimulus capable in the absence of the other? What is the relation of the two combined to each separately? These are questions which it is not possible to answer with any degree of certainty—yet a few words may be said with regard to them. There is no doubt about the status of the nervous mechanism as a definite stimulus. Apart from the older work of Richet (1887) upon polypnoea, the work of Krogh and Lindhard (1913) has proved conclusively that on the onset of exercise the increase of total ventilation is initiated in the nervous system and takes place before it is possible for the respiratory centre to be affected by any alteration in the hydrogen-ion concentration of the blood. Even if the subject (on a bicycle ergometer) thought he was taking exercise and was not in reality doing so, the total ventilation increased. The precise status of CO_2 as a "stimulus" in the restricted sense of the word is more doubtful. The reader can only be referred to Chapter 1 of this work and must be left to draw his own conclusion, firstly, as to whether he is sympathetic with the views there put forward, and secondly, as to what exactly is meant by the word stimulus.

When we ask: Of what is the nervous stimulus or the chemical stimulus capable, each without the other? the answer is not easy. Let us take the chemical stimulus first. It is of course easy to give CO_2 and equate the increase in the total ventilation against the pressure of oxygen in the alveolar air. But the nervous element is not excluded in this operation. CO_2 deepens the respirations, but each respiration in proportion as it deepens sends a discharge of impulses up the vagus which in turn modifies the respiration. A purer result would be obtained with the vagi cut, but even in this case afferent impulses from the stretching of the thoracic muscles and the diaphragm are not excluded. The experiment of measuring the total ventilation with the vagi cut can be carried out in some unanaesthetised animals.

The preservation of a level CO_2 pressure in the alveolar air whilst the depth and type of respiration is being varied independently by the nervous system presents an equally difficult problem. Whilst therefore we may cite the nervous and chemical factors in the regulation of respiration as an example of a duplication of mechanism for the attainment of a single end, it does not seem possible to regard them as entirely independent variables.

In the discussion to which I referred, one point made, or at least attempted, was that the nervous control of respiration was of a wholly more delicate character than the crude chemical control. I can recollect no suggested system of units by which the two were to be compared, but this I suppose is true: that not only is the play of afferent impulses on the respiratory centre capable of producing a great alteration in the total ventilation, but for any particular total ventilation it can produce a great variety in the ratio of frequency to amplitude, *i.e.* a great variety in the type of respiration. CO_2, on the other hand, for any total ventilation produces a respiration of a given type.

The matter did not strike me as a very important one then and does not do so now. I mention it rather because a similar problem will arise in the case of some other duplications of mechanism. It gravitates about the words "protopathic" and "epicritic" and it is not without interest that in the discussion quoted the point was raised by Henry Head.

Leaving aside the issue as to which is the finer and which the cruder mechanism, the important point to note is that though the two mechanisms, as a first approximation, achieve the same object, namely the alteration of the total ventilation, in detail they are different. Another question which rises rather naturally is the date of each mechanism in the evolution of the organism. Pulmonary respiration is itself a rather recent method of effecting gaseous exchange. In the more primitive forms of life, where do we first meet with a mechanism for the purpose of graduating the external medium of gaseous exchange to the needs of the organism? When we do meet that mechanism, is it chemical or nervous?

So far as the vertebrates are concerned, the researches of Lumsden (1923) seem to indicate that the chemical regulation of respiration, even if it is not a fundamental phenomenon, appears, though in a very restricted form, in the Chelonian.

If Lumsden is correct, if in the fishes there is no regulation by the carbonic acid of the water which flows through the gills, the fact is scarcely surprising. When the relative solubilities of carbonic acid and oxygen are considered, it is difficult to avoid the conclusion that the acquisition of oxygen must have prescribed a more difficult problem in the piscine economy than the discharge of carbonic acid. If that is so, one would look for the most primitive chemical mechanism for the regulation of respiration as one which centred about oxygen want rather than CO_2 excess, and, so far as we know, the influence of oxygen want is an indirect one. It is regarded by Haldane, and I think his view is generally accepted, as an alteration in the irritability of the respiratory centre. Indeed, though the effect of CO_2 on the respiratory centre may be traced down to the Chelonians, its great *development* is not a phenomenon which goes very far down the vertebrate world.

The terms which I have used constantly, "chemical stimulus" and "nervous stimulus", open up a much larger avenue of thought than is concerned merely with respiration. Perhaps the instance which first comes to the mind of a double mechanism of which one

member is chemical and the other nervous, is the duplicate control
of the vascular system by adrenaline and by the sympathetic nerves.
Yet as an example it requires some justification and that for two
reasons. In the first place adrenaline stimulates the nerve endings
of the sympathetic system, and in the second, the secretion of
adrenaline is itself a manifestation of the activity of the sympathetic
system. Then there is the more speculative point as to whether
the terminal reaction of every sympathetic ending is not really a
secretion of adrenaline or some closely allied substance.

If for a moment we may accept the doctrine of hormone secretion
—it is not to be thought that I am expressing any particular
opinion for or against it—the position would be that the central
nervous system can control the supply of adrenaline to the vessels of
a certain organ by a duplicate mechanism, either it can stimulate the
suprarenal body and the adrenaline liberated can find its way through
the general circulation to the point in question, or it can stimulate
the nerve endings in the vessels concerned. The duplication then
exists, but the alternative is not between a chemical and a nervous
control, for each mechanism is both nervous and chemical. The
duplication is one of path. If the direct nervous control failed the
central nervous system could effect some crude regulation via the
adrenal bodies; if the adrenal medulla were deficient, the vessels
could still be made to contract by their own proper supply.

Inadvertently, but I suppose naturally, the word "crude" slipped
off my pen. It raises the whole issue which arose over the regulation
of respiration and will rise over most of the duplicate mechanisms
which we have yet to consider—the issue of the relative delicacies
of the two units. The effect of liberation of adrenaline from the
suprarenal medulla must be a mass affair, it cannot control any
particular set of vessels without influencing all the nerve endings in
the body which are susceptible to it. Moreover, if the object were
to produce a certain flow through the vessels of, say, the kidney it is
evident that not only the concentration in the blood proper to that
particular degree of local contraction would have to be regulated,
but account would have to be taken of the secondary effects upon
the kidney of all other alterations in the vascular system (e.g. dilatation in the muscles) produced by the secretion of the adrenaline.

Whether one considers the grading of the stimulus, the degree of its localisation, or the rapidity of the response, it is clear that direct stimulation from the central nervous system must be much less "crude" than hormonal stimulation. From the moment the impulse leaves the brain, I imagine it would take no longer to produce a local constriction than to produce an injection of adrenaline into the blood, but once in the blood before the adrenaline takes effect it must needs pass through the heart, through the lesser circulation, through the heart once more, along the arteries to the capillaries, after which it must get absorbed, all lengthy processes as compared with the mere passage of an impulse down say three or four neurones. And when the system is once flooded with adrenaline, the endings on which it acts must respond so long as the adrenaline is present. The action of the endocrines, if less delicate than that of the nervous system, is more sustained, as may be shown by a comparison of the actions of a nervous stimulus and of injection of adrenaline and of pituitary extract. It needs no argument I think to establish the crudity of a general secretion of adrenaline considered as a regulator of the blood flow in this or that organ.

Another question which arises is the relative antiquity of the nervous and humoral control of organs in the evolution of the organism. The matter has been treated in general terms by Starling (1930) whose view, which is frequently quoted, is as follows:

Among the stimuli which, coming from a distance, evoke the reaction of unicellular organisms, probably the most important are chemical. These reactions are classed together under the term chemiotaxis. When the cells are united to form cell colonies or when, as in the metazoa, the multicellular aggregate is formed by the failure of the products of division of the ovum to separate one from another, the interrelations between the different cells of the organism are still largely determined by chemical stimuli. In fact, in the lowest metazoa such as the sponge, we know of no other means of correlating the reactions of different parts of the cell aggregate. The same chemical sensibility determines the aggregation of phagocytes around bacteria or dead issue, which forms the essential feature of the process of inflammation in higher animals.

When the reaction of distant parts of the body to a change occurring in any one part depends on the diffusion of some substance from the stimulated part, the total reaction must require a considerable time for its full development. A much more effective method of correlation was acquired by the evolution of a nervous system, by means of which *consensus partium* could be maintained by the rapid propagation of molecular changes along differentiated paths in the protoplasm. The development of this second mode of correlation of activities did not, however, do away with the necessity for the more primitive method. Even in the higher animals, when rapidity of action is not required, we find adaptations carried out in response to some change in distant parts of the body, the message having been chemical and not nervous in character (*e.g.* the secretin mechanism for pancreatic secretion).

In the present case, adrenaline is usually regarded as acting upon the terminal nerve endings of the sympathetic system. That thesis was never really admitted by Langley; nevertheless, many persons would say that the exceptions were such as proved the rule. It is difficult to see what would have been the rôle of adrenaline before the nerve endings were evolved. Surely the hormone control of the sympathetic system cannot have been evolved before the system which the adrenaline controls. I know of no animal which contains a supra-renal body but no nervous or even sympathetic system.

An obvious case of duplication is afforded by the alimentary canal—so obvious indeed that there is little to be said about it. It differs from the cases which we have discussed in that the members of the "duplicate" are not "in parallel" but "in series". At two quite different situations in the canal does both amylolytic and proteolytic digestion take place, and for each type of digestion two quite different organs are employed; the salivary glands and the pancreas for amylolytic digestion and for proteolytic digestion the stomach and the pancreas.

If one compares the properties of these "twins" with the cases already considered, one sees at once certain similarities. Just as in the case of the control of the vessels, one sees a nervous and a humoral form of stimulus. The amylolytic enzyme is secreted under the aegis of the nervous system from the salivary glands and under the influence both of the vagus and secretin from the pancreas; the proteolytic enzymes are secreted under the influence both of the nervous system and of hormones in the stomach and in the pancreas. While we may go so far, it becomes very difficult to go much further. Hitherto we have regarded the nervous function as implying delicacy and quickness, while the humoral function is rather crude, slow and massive. We scarcely know enough to pursue this comparison. It is, of course, clear that if the chorda tympani be stimulated, saliva will appear in four seconds, whereas if secretin be injected, perhaps 100 seconds or more elapse before the pancreatic juice commences to flow, but a differentiation into delicate and crude would demand something more far reaching— something one would expect, relative to the concentrations of the various components of the juice. Can the chorda tympani produce saliva containing a greater or less concentration of enzyme? Can secretin produce a pancreatic juice containing a greater concentration of trypsinogen, and in any case what is the nervous control of the pancreas and is there in life any humoral control of the sub-maxillary gland? Can the vagus produce a gastric secretion with a graduated concentration of pepsin? These are questions to which if there be an answer I do not know it.

Let us take these questions *seriatim*. Can the nervous mechanism of salivary secretion graduate the composition of the saliva?

It is of course certain from the work of Pawlow that the nature of the saliva in the dog is regulated in sympathy with the nature of the food taken. These differences are usually attributed to the fact that there are three salivary glands and that under different stimuli the mixed saliva is formed from the secretion of these glands in various proportions.

Yet there is some evidence that even from a single gland the nature of the saliva is not independent of nervous impulses as was shown by Heidenhain (1878) and Werther (1886) and Langley and Fletcher (1889).

Rate of secretion c.c. per min.	0·4	0·5	0·76	0·90	1·33
Percentage of solid	0·47	0·51	0·60	0·62	0·63

Similar results were obtained by Langley and Fletcher for saliva secreted as the result of stimulation with pilocarpine. It would look therefore as though the nervous system had some power of controlling the nature of the saliva. A clearer issue as to how this is done might have been obtained if the parotid, rather than the sub-maxillary had been used: for the parotid is a pure gland in which all the cells are alike, while the sub-maxillary is a mixed gland consisting of mucous and serous cells. It is worth remembering also, when trying to visualise what is taking place, that the saliva is less rich in solids than the plasma and that the data given above might be interpreted as meaning that the more rapid the flow of saliva, the less does the saliva differ from plasma in constitution.

Apart from these considerations, the fact remains that there is quite a number of ways consistent with known facts by which the central nervous system could graduate the quality of saliva, *e.g.* by marshalling the different glands in different degrees or again in the gland by intense stimulation of a small number of cells, or by mild stimulation of a large number.

Can the vagus produce a gastric secretion with a graduated concentration of pepsin? It is not very easy to answer the question precisely. Gastric juice secreted as the result of nervous stimuli certainly differs in composition, but the difference is doubtless due to the integrated effect of the nerve fibres which supply the peptic cells, the oxyntic cells and the blood vessels. From the last category the sympathetic is difficult to exclude. The relevant point however is that as the result of a purely nervous mechanism, whatever the fibres involved, it is possible to obtain gastric juice of composition appropriate to the character of the food. [The reader is referred to the article by Babkin (1928) in *Physiological Reviews*.]

So far as the pancreas is concerned, while there is evidence that the nature of the juice evoked by secretin is different from that resulting from nervous stimulation, the former being more copious and less rich in enzymes, there is no satisfactory evidence of specific alterations in the character of the juice evoked as the result of humoral stimulation.

An example of duplication which presents several points of interest is associated with the initiation of the heart beat. Normally originating in the sino-auricular node it is conducted down the bundle of His and spreads over the heart. If however the bundle ceases to conduct in its upper portion, a new rhythm originates lower down, and causes the ventricles to beat, but with a slower rhythm than before. The details of this mechanism are so well known that it is unnecessary to discuss them at any length. It may be argued in this case with even greater cogency than in those which have already been discussed that there are not two mechanisms but only one. Into this discussion I do not propose to go, for two reasons, firstly that most of it is self-evident and secondly that this solid fact remains: whether you are prepared to recognise two mechanisms or only one, the body has a solid insurance; if the normal method breaks down, another method is available; if the normal machinery is wrecked, the heart does not on that account cease to beat, as the beat is initiated elsewhere. And that is the point for which I am contending; fundamentally there are two mechanisms, if one breaks down the other can adequately though perhaps not efficiently "carry on".

In the instances of duplication hitherto discussed one of the members has been definitely more delicate in its action than the other. Can the nodal mechanism of the heart be described as crude with the normal one? Surely the answer is in the affirmative; not only is the rhythm slower, which perhaps is not in itself a sign of crudity, but the nodal beat is less capable of being attuned to the needs of the body.

The sympathetic always has some action on the nodal beat as is shown by experiments in which the bundle is crushed. The positive result with the vagus is much less certain—the experiment is not quite a clean one on account of the difficulty of crushing the bundle without injury to the nerve fibres, so that the result may be due to injury of the vagus fibres during the experiment. In any case, clinically, the amount of exercise which will raise the frequency of a normal rhythm will not raise the frequency of the nodal rhythm and sometimes even considerable exercise will not quicken a nodal pulse.

Of the arresting features of this instance of a double mechanism, one is the almost complete suppression of one member, the nodal rhythm. Always available, the great majority of mankind pass their whole lives and I suppose leave this world without one single nodal beat having taken place in their hearts. Perhaps one thousand million pulsations per person and for all but an insignificant portion of the population, never a single nodal beat, yet in the background it is there. This suppression is something which we have not met before: during the whole of life, the nervous and the chemical mechanisms for the regulation of respiration play into one another's hands and the mass secretion of adrenaline supplements the finer control from the nervous system. But, if we have not met suppression before we shall meet it again, though perhaps not in so extreme a form.

The other point which makes the nodal beat of especial interest is its place—if indeed it has a place—in the scheme of evolution. What is the "survival value" of the nodal beat? Clearly it secures the survival of a certain small number of persons, to whom it is a matter of life and death—but on the other hand the number is so small—too small one would suppose to secure the nodal beat a place as something which has been inherited by mankind because it prolongs the life of the race. Is it a happy accident? Are we to suppose that all of us carry an accident so happy in our hearts, that it will save the lives of a few persons whilst functionally it never enters the economy of more than a negligible proportion of the race? Are there such accidents? Are there accidents at all, or does the mill of evolution grind accidental mechanisms remorselessly out of the economy of the organism?

A dozen well attested instances such as that of the nodal beat, of something which in the overwhelming majority of mankind never even materialised but which also held the issue as between life and death to an insignificant number, would go far to make me think kindly of some form of teleology which did not depend upon survival values.

If we are to accept the views of the school represented by Head and Rivers, and more recently by Sir Herbert Parsons, the retina presents a very beautiful example of duplication—beautiful because in it are combined most of the characteristics which are found severally in the examples which we have hitherto cited, and therefore we may treat it in some detail as the "Compleat" instance of duplicate mechanism.

One of the best supported hypotheses relating to vision is the so-called duplicity theory. According to it scotopic vision or vision under low intensities of illumination by the dark adapted eye is carried out by means of the rods and photopic vision or vision under higher intensities of illumination by the light adapted eye is mediated by the cones. It would seem at first sight that we have here ready to hand the duality which is demanded by the dyscritic or epicritic theory. And indeed there are many facts which strongly support this view. The view put forward is that:

(1) Rod vision is cruder than cone vision inasmuch as cones see colour and rods only see light.

(2) The rods are phylogenetically a more ancient mechanism which has largely become superseded by the more recent and more exact vision of the cones.

(3) The rods though apparently capable of little but mere appreciation of light, and that up to no great intensity, have in reality had their functions suppressed, and in the absence of the cones would be capable of giving the organism much more information than normally they do.

(4) Occasions occur in which one form of vision is deficient, and in such cases the person is saved from blindness by the other.

With paragraphs (1) and (4) as statements of fact no one will disagree, but what basis is there for paragraphs (2) and (3)? Is the rod system phylogenetically the more ancient and primitive? Is there sufficient reason for supposing that in the process of its replacement by the more delicate cone system, the functions of the rod system have become suppressed?

There is overwhelming proof, derived from peripheral luminosity curves, minimal field and minimal time curves (see Parsons, *Colour Vision*, pp. 76–8), that peripheral vision behaves in exactly the same manner as central vision but with diminished sensitivity. Greater stimuli are required to produce equivalent responses but if the stimuli are sufficiently great the differences disappear, including even qualitative differences so that the fields for colours extend to the extreme periphery. Anatomically the peripheral neuro-epithelium consists almost entirely of rods—the scotopic or dyscritic mechanism. (Parsons, 1927.)

And now as to the relative antiquity of the rod system.

Evidence of the antiquity of the "rod system" is far from complete. Let us at the start break the conception of the "rod system" into two: (1) the rod and (2) the system, or in greater detail (1) the anatomical or histological structure known as the rod including the chemical material known as visual purple, and (2) the type of vision which the rod subserves. The weak point in the evidence of the antiquity of rod vision is that such evidence concerns itself almost entirely with the second division of the subject. To some minds that may not be a weak point, for it is clear that physiological function in the central nervous system is becoming less attached to anatomical structure than was formerly supposed. Nevertheless, to make a complete story one would like to find in the most primitive eyes rods without cones and visual purple without any other medium for picking up light. So far as I know there is no evidence of this. There is no reason to suppose that the cone and whatever it represents chemically is a later development and the rod an early one.

The evidence of the antiquity of the rod system rests largely in the use of the word "primitive" and it is not quite clear that the word is always used in quite the same sense—it is one thing to be primitive functionally and another to be primitive phylogenetically. I suppose it is true that functionally rod vision is less separated than cone vision from the most primitive sensation appreciated by an organ which is definitely an eye. That sensation is probably little more than consciousness of the movement of objects in the vicinity of the sense organ.

However, the relative antiquity of the rod and cone systems is not the primary question in our present discussion. The subject is that of duplication and, regardless of phylogenetic interests, the fact remains that in the human eye there are two systems, that though they subserve somewhat different functions in detail, those functions overlap and therefore if one or other function is lacking, as is sometimes the case, vision of a kind persists.

And so we come to the whole question of cutaneous sensibility, for the view of retinal vision put forward by Parsons is only a special case of the larger division of sensibility into protopathic and epicritic. Parsons uses the word "dyscritic" where Head and Rivers (1908) used "protopathic", but functionally the idea is the same—that for the appreciation of sensation there are two systems, one crude, of ancient development and normally suppressed, the other recent and capable of the appreciation of finer gradations of sensation. Indeed in the extreme view protopathic sensations were supposed to be incapable of gradation. They were supposed to be merely qualitative, if I may be pardoned for describing in crude language sensations of which crudity was one of the leading alleged characteristics. Head and Rivers and also Parsons invoked the "all-or-none" principle to describe the phenomenon, holding that protopathic sensations were in accordance with that principle and epicritic sensations were outside it.

I omit any emphasis on the "all-or-none" principle, because the more recent work on the nature of sensation suggests that the impulses which traverse one sensory nerve fibre "are uniform in character, and differ only in frequency". I suppose to-day if a particular sense organ were said to be capable of yielding only qualitative information, the implication would be that somewhere in the path the power of appreciating the frequency of "all-or-none" stimuli was lost.

The principle put forward may therefore still hold good. The skin of the prepuce may be capable of giving quantitative information to the brain concerning touch or pain while that of the penis may only be capable of giving qualitative information and the difference may be important, but in the present state of knowledge one would suppose it to be a difference somewhere in the power of integrating "all-or-none" responses and not involving any question of the "all-or-noneness" of the impulses themselves.

In the above discussion of the rods and cones four points were cited as being claimed for the protopathic type of vision, (1) crudeness, (2) antiquity, (3) suppression, and (4) an emergency value. What have the authors of the protopathic and epicritic view of cutaneous sensation to say about these four points?

(1) "Prolonged observations after the division of the nerves in Head's own arm brought out clearly the existence of two definite stages in the return of sensibility. In one of these, the protopathic stage, the sensations are vague and crude in character, with the absence of any exactness in discrimination or localisation and with a pronounced feeling-tone usually on the unpleasant side, and tending to lead explosively, as if reflexly, to such movements as would withdraw the stimulated part from contact with any object to which the sensory changes are due. At this stage of the healing of the reunited nerve there is present none of those characteristics of sensation by which we recognise the nature of an object in contact with the body. The sensations are such as would enable one to

know that something is there and is pleasant or unpleasant. It is also possible to distinguish between mere contact or pressure and stimulation by heat or cold, but within each of these modes of sensation there is no power of distinguishing differences of intensity nor of telling with any exactness that spot where processes underlying the sensory changes are in action." (Rivers, 1920 a.)

(2) "In interpreting these observations two chief possibilities are open. Epicritic sensibility may be only a greater perfection of protopathic sensibility, experience gradually enabling an exactness of discrimination and localisation which was not at first possible. The other alternative is that the two kinds of sensibility represent two distinct stages in the development of the afferent system. According to this second view" (the one of course which Rivers accepts) "the special conditions of the experiment revealed in the individual two widely different stages in the evolution of cutaneous sensibility.... The way in which epicritic sensibility returns and the fact that it is possible to annul it by treatment affecting only the peripheral factors without influence on such central processes as would be set up by experience, go far to show that the two modes of sensibility represent two stages in phylogenetic development." (Rivers, 1920 b.)

(3) "When epicritic sensibility returns, the earlier protopathic sensibility does not persist unaltered side by side with the later development, but undergoes certain definite modifications. Some of the elements persist and combine with elements of the epicritic stage to form features of normal cutaneous sensibility. Thus the cold and heat of the protopathic stage blend with the modes of temperature sensibility proper to the epicritic stage, and form the graded series of temperature sensations which we are normally able to discriminate. The crude touch of the protopathic system blends with the more delicate sensibility of this kind, while protopathic pain with its peculiarly uncomfortable rather than acute quality forms a much larger element in the normal sensibility of pain. In the process of blending or fusion certain aspects of the earlier forms of sensibility are modified to a greater or less extent, and in some cases this modification involves the disappearance of certain characters. This disappearance is especially striking and complete in the case of spatial attributes of protopathic sensibility. In the

protopathic stage (where the sensibility of the deeper structure is excluded) there is no power of exact localisation. When a point of the skin is stimulated the sensation radiates widely and is often localised at a considerable distance from the actual place of stimulation. These two characters of radiation and distant reference disappear with the return of epicritic sensibility and afford examples of a process of suppression....Moreover, these spatial features of protopathic sensibility do not disappear entirely but persist in a latent form ready to come again into consciousness if the appropriate conditions are present." (Rivers, 1920c.) "The appropriate conditions" being the loss of epicritic sensibility, we see in the author's statements, if we accept his reasoning, evidence not only of suppression but of insurance.

(4) Let us pass to the emergency value of protopathic sensation. The reader may wonder why I instanced the rods and cones before discussing cutaneous sensation when, in point of fact, Parsons's classification of retinal function was built up upon the work of Head and Rivers. The reason was as follows. We are discussing duplication and clearly there are two visibly separate elements in the retina, the rods and the cones. We start therefore with a basis of fact which is beyond dispute. Discussion can only, though it very legitimately may, centre around the question of whether Parsons has correctly interpreted their functions. But with cutaneous sensation it is different; there are many sorts of sensory end-organs in the skin. Is it possible to sort them into those which are epicritic in function and those which are protopathic?

The problem is one which has fascinated more than one author and the work upon it can only be surveyed very briefly.

There are parts of the body which, if touched, are credited with the ability to feel crude protopathic sensations of awareness of a painful character and are denied the power to experience epicritic sensation of any kind. Such are the cornea and the epithelial covering of the glans penis. In the cornea the only form of sensory organ present is the plexus of nerve filaments in the epithelial covering. So far as I understand the sort of primitive sensation which the word "protopathic" is intended to denote, it is as nearly as may be that which these two organs possess. If someone touches your cornea and if you feel the touch at all you experience a sort of massive pain, not accurately located nor accurately assayed, but which is calculated to make you hit out. [A case came within my own knowledge. An officer was in charge of the swimming team of his regiment. In the swimming bath one of the men accidentally kicked the said officer hard in the eye, the response was immediate, the officer being muscularly very powerful dealt the man such a blow as sent him to hospital. The action violated all the habitual instincts of affection and regard which the officer had for his men, as well as (I suppose) violating military discipline. Whether the matter came to a court martial I know not, but had it done so and had Henry Head been called as a witness, I doubt not that he would have represented the blow as a quite inevitable response to a primitive stimulus: in fact the emergency value of protopathic sensation.]

There are other places in which the only form of sense organ is the naked terminal, such as the blood vessels, and from them the only obtainable conscious sensation is pain.* The pain is of the same character as that felt in the cornea—as I know from the experience of many arterial punctures. I know also that the reaction is one which prompts violence.

Having dealt with the above four points, and before considering criticism of the epicritic-protopathic theory of cutaneous sensation, let me pass to a subject which has been mentioned incidentally, namely, the distinction between encapsulated and naked endings. This difference has been stressed by various workers on the subject but from different points of view of which the following two are the principal.

(1) There is a difference in physiological properties between normal nerve endings of the naked and encapsulated types.

Of the various cutaneous nerve endings those in the cornea, the penis, the epidermis of the skin, etc., fall into one group, the naked

* Adrian has put forward the suggestion, that the normal response in such cases is a subconscious one, in the case of arteries a response to pressure such as is most evident in the carotid sinus and in the case of the teeth it is the sort of sensory stimulus which, if present, produces a trophic response, and without which the teeth deteriorate. Only when such sensory stimuli are exaggerated do they express themselves as pain.

type, whilst the touch corpuscles and the like fall into the group of encapsulated organs.

The neurilemma and the end organ of the peripheral sensory nerve are regarded as having important insulatory functions. It is suggested that if the free terminal ramifications of sensory nerves are the end-organs subserving pain sensibility, the specific quality of the sensations resulting from stimulation of them may be due to a slight normal irritation consequent on incomplete insulation. The suggestion is strengthened by the observation of *von Frey* that for faradic stimulation the pain threshold of the skin is lower than the touch threshold. (Trotter and Davies, 1913 a.)

Without attaching too much importance to the capsule as possessing any particular function, it does at least imply a definite boundary to the organ. For instance, if a touch corpuscle be compared to the terminal net work in the cornea, each touch corpuscle is a discrete entity, but I gather that though much work has been done on this subject no one can say quite definitely whether the filaments say in the cornea do or do not anastomose—they are drawn as doing so (Maximow, 1930; Schafer, 1920)—or if they are not continuous, whether they are contiguous in the sense that an impulse can pass from one to the next. In either case there is the possibility that a stimulus applied to a spot on the cornea would give a less localised sensation than one applied to a spot on the skin. Even if the fibrils never actually touch there is still the possibility of their wandering about in such a way that areas supplied by the terminals of various fibres may overlap and that to a great degree.

It is reasonable to suppose that the insulation of other end-organs (the roots of the tactile hairs, the various forms of touch corpuscle, the Pacinian bodies and the muscle spindles) does suggest a more definite degree of localisation and possibly of gradation. It is intelligible that compass points, if they fall on different touch corpuscles should be appreciated as being two in number, whilst if they fall on different portions of a plexus, they should only be appreciated as "something".

If then we regard the various forms of encapsulated end-organ as differing from one another by no more than mere differentiation in detail, and if further we classify end-organs into those which are naked and those which are encapsulated, we are faced with the question: Is that detail too? Are there really two kinds of end-organ, or is there only one? I am not morphologist enough to give an answer with any authority, but so far as my personal bias goes I incline to there being two, but like the rods and cones I suppose they have been evolved from a single type, and while the naked organ is objectively simpler I agree with Schafer that there is no reason to regard the sensation of pain as being more primitive than that of touch.

(2) The distinction between the properties of naked and insulated nerve fibrils takes on another aspect and one which has been much stressed by Trotter and Davies (1913), namely, that the regenerating fibre proper to an encapsulated organ may have different properties while it is growing to meet the organ than it has after it has made contact, or rather different properties from the complete unit, fibre plus encapsulated organ.

Pathological evidence tends to show that excessive response to stimuli is an indication of persistent slight irritation of the nerve. It is therefore suggested that the intensification of sensation elicited from the recovering area is due to the regenerating nerve fibres being subject to chronic irritation.

The source of this irritation is found in the contact of the new nervous tissue with the non-nervous tissues of the part and the consequent reactions set up. (Trotter and Davies, 1913 a.)

Perhaps we have gone far enough in the domain of nerve terminals; let us turn in the other direction to the spinal cord.

I will be brief about the sensory tracts in the central nervous system, and for several reasons. The first of these is the sense of my own incompetence—a second is the fact that the question is treated in so pleasant a way by Stopford (1930) in his book *Sensation and the Sensory Pathway*. So far as I follow him, the position as regards pain is as follows: all sensations of pain are conducted by fibres which enter by the posterior roots and end within about three segments of their points of entry. There are then two possibilities of conduction up the cord.

(1) From segment to segment, a form of conduction much in evidence in the lower animals, but which has dropped out almost entirely in man.

(2) By the spino-thalamic tract.

Sensations of temperature of all kinds enter by the posterior root, cross over almost immediately to go up by the spino-thalamic tracts.

Tactile sensations do not cross so soon; some pass by fibres which enter the grey matter before reaching the medulla; near the point of entry there is a synapse and the second relay crosses, and passes upwards in the anterior spino-thalamic tract. The remaining sensations of touch are such as discriminate the distance apart of two compass points, these go up to the nuclei gracilis and cuneatus of the same side whence the impulses are relayed to the fillet of the opposite side.

All the fibres of conscious sensation have now crossed. As a parenthesis we may mention paths of unconscious sensation which have to do with posture, movement and equilibration. Here again we have duplication, for some of the impulses go directly to Clark's column and up the direct cerebellar tract, whilst others go up in the columns of Goll and Burdach, all passing to the restiform body.

In reviewing the evidence with regard to various forms of protopathic and epicritic sensation I do not find it easy to see a uniform picture.

Let us tabulate the structures concerned:

Table XXXIX

	Protopathic	Epicritic
Vision	Rods	Cones
Tactile sense	Naked nerve ending	Encapsulated endings
Tracts in cord	Spino-thalamic	Posterior column
		Spino-thalamic
Ultimate destination	Thalamus	Cerebrum

Now let us revert to the criteria of protopathic as opposed to epicritic action.

The spino-thalamic tracts are, I believe, more ancient than the posterior columns and the thalamus more ancient than the cerebrum; but I know of no real reason for supposing that the rods are more ancient than the cones.

Head and Rivers performed but a single experiment; on that foundation they built a theory of the peripheral phenomena of cutaneous sensation. But the building only commenced there; another storey was added by Parsons who has applied the theory to visual sensation, a third storey has been put on by Stopford in his general application of the epicritic and protopathic theory to cerebral and thalamic phenomena, while finally Rivers has crowned the whole with a system of psychology and treatment. Did ever so small a foundation carry an edifice of such magnitude? And if the foundation is faulty, what is to happen to the building? Quite frankly, at the expense of being thought an iconoclast, I cannot escape the belief that the last word has not been said with regard to the relation between the cutting of a nerve trunk which supplies an area of skin and the complete denervation of the area in question. The different authors give such different accounts of what takes place immediately after the section, *i.e.* in the first week or so. I append three statements on this subject: (1) by Schafer (1929) concerning an area which he regards as completely denervated; (2) by the same author concerning one in which there was some degree of innervation from sources other than the cut nerve trunk; (3) a résumé by Stopford of Head and Rivers' result. The reader could wish for a reliable histological report on each of these cases.

(1) *Schafer "denervated"*. "As soon as the pain caused in the above manner had disappeared" (*i.e.* a continuous pain, referred from the cut stump and "not provoked or altered by excitation within the area") "...cutaneous sensibility of every kind was abolished."

(2) *Schafer "incompletely denervated"*. "The result of section was to produce hardly appreciable loss of sensibility. Pain sensation, warmth, cold and touch were all felt; touch was slightly diminished, but so slightly that it was not easy to be sure of the fact."

(3) *Stopford's résumé of Head and Rivers*. "Shortly after the operation performed by Sherren, a rapid examination of the hand revealed the fact that the pressure of any blunt instrument was readily appreciated everywhere, in fact it was stated at the time that sensory disturbance was such that rough clinical tests commonly employed would have failed to reveal any defect of sensation

and the abnormal state to be described might have been overlooked by anyone unfamiliar with refined modes of examination."

It is the unwritten law of scientific experiment that it should be repeatable. The more difficult the experiment and the more complicated its interpretation, the more necessary does repetition become. Head and Rivers did not repeat their experiment. Others have done so, notably Trotter and Davies (1913) at University College Hospital, Boring (1916) in New York, and Schafer in Edinburgh. In each research several areas were denervated and under different conditions. These authors agree that, even with Head and Rivers' records available, they would have hesitated to place any final interpretation on their own first efforts. They agree also that the theory of protopathic and epicritic nerves lacks experimental basis. Schafer says:

> Although they (*i.e.* Trotter and Davies) observed certain of the phenomena described by Head and Rivers, they were led to adopt an entirely different explanation, and the criticism to which they subjected Head's theory was so damaging, and backed by such conclusive arguments, that one is surprised that Head's theory and terminology are still widely accepted....

Here then we must leave the question of protopathic and epicritic sensation until someone can be found who firstly repeats Head and Rivers' experiments and secondly confirms their results. But if Head and Rivers' theory goes, what remains? What then is life, if the whole structure of epicritic and protopathic sensibility comes tumbling down? Just this, that the skin, like the eye, has more than one mechanism for the appreciation of contact with foreign bodies; it has three, namely, temperature, touch and pain (in temperature the contact is incidental); they have different nerve endings, they pass by different paths and they acquaint the person of contact by different sensations.

In each case there appears to be a definite duplication. If the cones ceased to function you could see something with rods, though not precisely what you see with the cones; if the encapsulated tactile organs were abolished you could still feel something with the naked terminals; if the posterior columns are destroyed (and the same is true of the appropriate regions in the cerebrum) you are not left entirely without the appreciation of the external world by the sense of touch.

And what is the case with sensation appears to be true of duplication of function generally—it is of frequent occurrence but there is no uniformity as to the method of its happening. So frequent indeed is duplication that it can scarcely be accidental; it seems to be a definite feature in the architecture of function, the more impressive because it is achieved in such different ways—but I have often been surprised that the heart itself was not duplicated.

It is always surprising to me that there is only one heart!

REFERENCES

BABKIN, B. P. (1928). *Physiol. Rev.* **8**, 365.

BORING, E. G. (1916). *Q.J. Exp. Physiol.* **10**, 1.

HALDANE, J. S. and PRIESTLEY, J. G. (1905). *J. Physiol.* **32**, 225.

HEAD, H. and RIVERS, W. H. R. (1908). *Brain*, **31**, 323.

HEIDENHAIN, R. (1878). *Pflügers Arch.* **17**, 1.

KROGH, A. and LINDHARD, J. (1913). *J. Physiol.* **47**, 112.

LANGLEY, J. N. and FLETCHER, H. M. (1889). *Phil. Trans. Roy. Soc.* B, **180**, 109.

LUMSDEN, T. (1923). *J. Physiol.* **57**, 153, 354; **58**, 81, 111, 259.

MAXIMOW, A. A. (1930). *A Text-book of Histology*, ed. W. Bloom, Fig. 204, p. 279. Philadelphia.

PARSONS, Sir J. H. (1927). *An Introduction to the Theory of Perception*, p. 177. Cambridge.

RICHET, C. (1887). *C.R. Soc. Biol.* Series VIII, vol. 4, 482.

RIVERS, W. H. R. (1920a). *Instinct and the unconscious*, p. 22. Cambridge.

—— (1920b). *Ibid.* p. 23.

—— (1920c). *Ibid.* p. 24.

SCHAFER, E. A. (1920). *Essentials of Histology*, Fig. 274, p. 199. 11th ed. London.

—— (1929). *Q.J. Exp. Physiol.* **19**, 103.

STARLING, E. H. (1930). *Principles of Human Physiology*, p. 949. 5th ed. London.

STOPFORD, J. B. S. (1930). *Sensation and the Sensory Pathway*. London.

TROTTER, W. and DAVIES, H. M. (1913). *J. Psych. Neurol.* **20**, 102.

—— —— (1913a). *Ibid.* **20**, 141.

WERTHER, M. (1886). *Pflügers Arch.* **38**, 293.

CHAPTER XV

THE CHANCE THAT A PHENOMENON HAS
A SIGNIFICANCE

Of the "features" which have been discussed in the foregoing pages the majority have been well defined; it is perhaps in the natural order of things that we should close with the discussion of something much more problematical. It must have fallen to the lot of even the least inquisitive, observing some phenomenon in the body, to ask himself "What is the significance of this?" He may then be conscious of a skeleton in the deductive cupboard. "Perhaps after all it has no significance: it may be merely an accident." If I may change the metaphor, he may view the path of intellectual enquiry as a crooked road from which branch not only paths each of which leads to a definite objective, but also *culs de sac*, every one ending blindly round the corner. The philosophy which countenances, perhaps I should say welcomes meaningless phenomena seems to me to be a modern growth and to be a reaction largely against two former lines of thought.

Seventy years ago, when a structure was regarded as a manifestation of Design, when man was held to have been made in some material sense in the Image of his Creator, the idea of accidental phenomena, of structures which had no function but simply which "were there", was practically ruled out. Such an idea would have been insulting to the One who was all wise.

Again in the last twenty years of the nineteenth century the air was full of statements about protective mechanisms, and more particularly protective colouration, which went far beyond anything warranted by the facts. These explanations, so far as protective colouration was concerned, depended upon the assumption, entirely untrue, that the predatory animals see the world with the same colour sense as we do, and therefore that two objects which to us look very much alike will do so to fish or beetles or what not.

To illustrate the untruth of this assumption it is unnecessary to go beyond the properties of the human eye itself. Man has two mechanisms for vision, one which functions in ordinary light, the other for dim intensities of light: but the former responds to wavelengths lower than those which act as appreciable stimuli to the latter. A very beautiful demonstration of this to an audience consists of a pair of paper silhouettes of some object—say a fish—cut from bright red and bright blue paper respectively. They are placed behind a curtain. At the appropriate time the room is rendered dark* and the lecturer discourses a little while the eyes of the audience accommodate to the dim light; then the curtain is withdrawn and they see only one fish—the blue one. Then the lights are turned on and the red fish becomes as conspicuous as its blue companion. The moral is that a race which had only our ordinary "daylight" mechanism for vision would find it very difficult to argue about what would or would not constitute protective colouration for an eye whose sole machinery for vision was visual purple or *vice versa*.

PULMONARY VASOMOTOR SUPPLY

Among the phenomena which have been alleged to have no significance is the neuromuscular armament of the pulmonary arterioles. The matter has never been stated precisely in that way so far as I know. Nevertheless, the persistent denials of vasomotor effects in the lung really amount to a plea for a phenomenon without a significance. No one denies either that the pulmonary arterioles possess muscular walls or that these walls possess nerve endings.

The conception of vasomotor nerves to the lung was put forward on an experimental basis by Bradford and Dean (1894). It is true that several, notably Brown-Séquard (1872), had advocated the idea of vasomotor nerves in the lung. But Bradford and Dean claim to have stimulated specific nerves and obtained results which they

---

* Such light as there is must be a white light, not ordinary artificial light. I have never tested a daylight "lamp".

regarded as attributable only to constriction of the vessels in the lung. They say:

> It is difficult to come to any other conclusion, except that the pulmonary effects produced on excitation of these roots are dependent upon an active constriction of the lung vessels, and not upon any indirect effect transmitted from the systemic to the pulmonary circulation.
>
> We have shown that a rise of pressure in the pulmonary artery can be produced by the excitation of efferent fibres.
>
> A. When cardiac acceleration is absent;
>
> B. When there is no change of pressure in the systemic vessels, *e.g.* the 3rd and 4th dorsal with or without previous section of the splanchnics;
>
> C. When the systemic pressure falls, *e.g.* 3rd dorsal;
>
> D. When the rise in pulmonary pressure is accompanied by a systemic rise, the former is out of proportion to the latter, *e.g.* the 5th dorsal;
>
> and finally that far greater rises of systemic pressure than any produced by the excitation of these nerve roots may be observed with synchronous smaller effects on the pulmonary pressure than those described above....

There can be no doubt either about the difficulty of this field of research or the care with which Bradford and Dean's work was done.

The difficulty of the research, as compared with similar observations on a small organ such as the salivary gland, lies in the fact that in the small organ whatever be the condition of the vessels in the organ, a relatively constant blood pressure can be maintained and an unlimited supply of blood available—to prove an alteration in the calibre of the vessel it is only necessary to measure the variations in venous outflow; but in the case of the lung the blood flow except over short intervals of time is dictated by the systemic circulation and perhaps, though this was regarded as more important when Bradford and Dean wrote than it is to-day, by the activity of the heart itself—all this quite apart from the inherent operative difficulties which are presented by the pulmonary circulation.

The work of Bradford and Dean was I think accepted, at all events it was not made the subject of adverse criticism by so severe a critic and so great an authority on the sympathetic system as Langley (1900). But in 1904 the work of Bradford and Dean was challenged. New perfusion methods had been elaborated. In order to maintain a constant pressure head of blood and an unlimited supply, Brodie and Dixon (1904) perfused the lung and failed to obtain evidence of vaso-constriction on stimulation of the nerve roots of the upper thoracic nerves. Moreover, they applied what was then a quite new method of analysis, namely the use of adrenaline. Arguing that if (1) the constrictor fibres came from the sympathetic system, and (2) all tissues innervated by the sympathetic system responded to adrenaline, then if there were vaso-constrictor nerves as Bradford and Dean asserted, they concluded that these additions of adrenaline to the blood used in the perfusion should constrict the vessels, but no constriction occurred. On the other hand they sometimes obtained dilatation. Brodie and Dixon did not stress the existence of dilator nerves, which were rather under a cloud at the time, and the reading public, as is frequently the case, accepted the major thesis of the authors whilst they neglected the qualifications with which this was safeguarded. The existence of vasomotor nerves to the lung was freely denied: that denial involved the conception that the neuro-muscular mechanism so easily demonstrated was a functionless phenomenon.

Much water has flowed under the bridge since those days. It is now possible to consider the subject comparatively.

In 1910 Krogh, in criticising Bohr's (1909) view that the vagus in the tortoise was responsible for the secretion of gas by the pulmonary epithelium, demonstrated that Bohr's results were to be explained by the vaso-constrictor properties of the nerve in question.

Important work on cold-blooded animals has been done by Nisimaru (1923).

(i) In the frog he cut the cardiac branches of the vagus retaining the pulmonary branches intact. He mounted the lung and the web of the foot under microscopes so that each could be observed. On stimulation of the distal end of the cut vagus there was slowing of the blood stream in the lung but not in the web.

(ii) The lung of the toad (*Bufo japonica*, weight about 300 gm.), when perfused, gave the following results (Table XL).

(iii) The information obtained about the tortoise may best be understood from Fig. 105.

Table XL

Procedure	Effect
1 c.c. of 1/5000 adrenaline added to Ringer's solution	Little effect
1 c.c. of 1/1000 adrenaline added to Ringer's solution	Marked slowing
Stimulation of vagus	Slowing
Stimulation of vagus after atropine	No effect
Wash out atropine	Slowing

1. Direct stimulation of the vagus produces slight slowing in the perfused lung.

2. Stimulation of the sympathetic trunk also produces slight slowing of the flow of fluid.

3. If the sympathetic be cut between Y and Z stimulation between Y and the ganglion of the vagus will produce constricting, but stimulation below the point of section will not do so.

The suggestion is that the constrictor fibres are sympathetic in origin, that like the accelerator fibres of the heart they pass towards the head, and join the vagus at the vagus ganglion and pass down in the mixed trunk.

(iv) Lastly to revert to the toad in which the roots are easy to stimulate, the II and III spinal roots appear to be those most deeply involved.

Plumier (1904) and Wiggers (1909), using perfusion methods, and Fühner and Starling (1913), using the heart

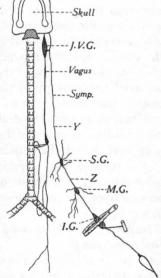

Fig. 105. Nerve to lung in *Clemmys japonica*.

lung preparation, showed evidence of vaso-constriction with adrenaline, and Argyll Campbell (1911) obtained it also with hemisine. All these observers agree with Bradford and Dean in the statement that the vaso-constriction is small relative to what may be obtained in other organs.

Whether there are vaso-constrictor fibres or not, the conclusions which Brodie and Dixon drew from their undoubted and accurate observations seem to be swept away by the work of Mrs Tribe [now Mrs Oppenheimer] (1914). I never understand why more stress is not laid on this paper. The work was really the direct continuation of Brodie's, done much of it, I think, with apparatus which Brodie left at the Women's School of Medicine on his departure to Toronto.

Mrs Tribe had no difficulty in obtaining Brodie and Dixon's results by following their procedure, but she also showed that adrenaline could produce vaso-constriction. Their procedure was modified in certain ways. These modifications were:

(1) The perfusing blood was not allowed to come in contact with the metal pump.

(2) The temperature of the animal was maintained by keeping it in a bath of 9 per cent. saline at 37° C.

(3) The preparation of adrenaline said to have been used most frequently—namely Parke, Davis' adrenaline hydrochloride—was avoided; that used was either the base or the hydrochloride prepared freshly from the base.

(4) Rather strong solutions of adrenaline were used.

So far as vaso-constriction is concerned, Mrs Tribe arrived at the same conclusion as Bradford and Dean, namely that the vessels of the lung were under the influence of vaso-constrictor fibres from the sympathetic to some slight degree, but only to some slight degree. My experience of perfused organs has been that a perfused organ tends to present a higher resistance than the same organ normally in the body and therefore I think it not unlikely that as compared with Bradford and Dean later workers handicapped themselves by working on preparations the vessels of which were initially in some degree of spasm. But even so there are some general considerations which suggest that all these observers worked at a disadvantage. These considerations are as follows:

(1) It would be highly undesirable that the vessels of the lung should be capable of so great a constriction as to produce actual obliteration of the lumen. In the submaxillary gland of the dog it is possible to produce an approximate stasis by stimulation of

the sympathetic. If to the possibilities of bronchial spasm were added those of vascular spasm the organism might have its expectation of life considerably curtailed. It may be highly desirable to deprive an organ in the systemic circulation of blood in order that the flow may be diverted to some collateral path—but it is not so with the lung as a whole.

(2) On the other hand, the more the animal is at rest the less is the flow of blood through the lung and the more the average calibre of the lung vessels is likely to approximate to its maximum degree of constriction.

It may be then that all these researchers were working on a rather narrow margin; that choosing as a subject for experiment a lung whose vessels were already much constricted and were incapable of complete spasm, Brodie and Dixon showed plainly enough that the vessels of the lung were capable of a considerable degree of dilatation. I should like to have seen the effect of a large dose of adrenaline on vessels known to be dilated. If I am pressed as to how the lung vessels came to be in a state of considerable constriction after the constrictor roots had been cut on one side—apart from the retort that they were not cut on the other—I can only take refuge behind my own ignorance. It is admitted generally that some chemical substance or substances are responsible for the maintenance of a basal degree of muscular tone; it seems not to be asking much of the evolution of unstriped muscle to postulate that the fibres in some situations are more sensitive to such influence than those in others, and that among the more sensitive are those of the pulmonary vessels.

The most recent workers in this field, with whom I have come in contact, are Daly and his colleagues. Daly has kindly given me permission to publish the following note:

It appeared to me, however, that the failure of Brodie and Dixon, and of Dixon and Hoyle, to obtain evidence of active pulmonary vasomotor nerves using the *isolated perfused lungs* must have some explanation. During the last four years we have been at the problem and in short this is the story.

Berry, Brailsford and I found that the pulmonary nerves received a rich supply of blood from the bronchial arteries and we thought that the failure of Dixon, Brodie and Hoyle to find an active vasomotor system in the lungs might be due to the death of the nerves owing to their lack of blood supply. Berry and I therefore devised an isolated lung perfusion preparation in which we perfused simultaneously the bronchial and pulmonary vascular systems.* Von Euler and I then set to work on this preparation and we were able to demonstrate the presence of active vasomotor nerves to the lungs. I think the tracing will support this contention (Fig. 106).

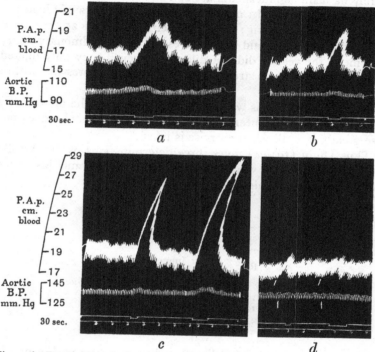

Fig. 106. Dog 9·0 kg. Isolated perfused lungs. Simultaneous perfusion of the pulmonary and bronchial (through the aorta) vascular systems. *a.* Stimulation of both stellate ganglia. Coil distance, 11 cm. *b.* Stimulation of both cervical vagosympathetic nerves. Coil distance, 9 cm. Between *b* and *c* 2·0 mg. of atropine sulphate was injected into the aortic circulation and 3·0 mg. into the pulmonary circulation. *c.* Stimulation of both thoracic vagosympathetic nerves. Between *c* and *d* 2·5 mg. of ergotamine tartrate was injected into the pulmonary and 1·5 mg. into the aortic circulation. *d.* Two successive stimulations of both thoracic vagosympathetic nerves. Coil distances, 14 and 5 cm. Primary voltage, 2·3 during each stimulation. *P.A.p.*, pulmonary arterial pressure.

* The preparation was under negative pressure ventilation with 95 per cent. $O_2 + 5$ per cent. CO_2.

These results have been published (Daly and von Euler, 1932) since the above was written.

While, however, I agree that the musculature of the pulmonary vessels has a significance from the point of view of constriction, surely its main significance must be to confer on these vessels the power of dilatation. In man at rest perhaps 4 litres of blood per minute traverse the lung; during full activity perhaps 25 or 30 litres per minute. What would not the pressure in the pulmonary artery become if the vessels did not dilate? or even if they only dilated in so far as they were stretched by the increase of pressure within them?

Yet such evidence as we have goes to show that the pressure on the right side of the heart, either in systole or diastole, does not rise greatly if at all when exercise is taken.

Thus Dunn (1919 a) gives three determinations of the systolic pressure in the right ventricle of the goat before and just after such exercise as doubled the blood flow.

Table XLI

Systolic pressure in mm. of Hg.	
Rest	Immediately after exercise
17	23·5
14	16·5
27·5	23

These were obtained by direct cardiac puncture with a needle attached to a tambour.

Dunn (1919 b) obtained results which were equally to the point as the result of multiple pulmonary embolism produced either by injection of starch grains into the right side of the heart (or jugular vein) or by the injection of olive oil.

The immediate result of the embolism was always to raise the right systolic pressure which may be taken as the pressure in the pulmonary artery, but this was only transitory and very soon the pressure settled down to the old level, e.g.

Table XLII

	Average diastolic pressure mm. Hg	Average systolic pressure mm. Hg
Before injection	5·5	18·0
After first injection, 2 c.c. oil	6·5	18·0
After second injection, 3 c.c. oil	14·0	23·0
10 min. later	4·0	17·0

As it did not appear that there was any certain diminution in the minute volume of blood, it can only be concluded that vessels opened up after the injection and compensated for those which had been blocked up by the emboli. The starch emboli were particularly easily seen on account of their staining properties, and it was evident that a very considerable part of the whole vascular area of the lung had been intercepted.

The subject may be approached in another way. In Chapter VIII we have referred to the diffusion coefficient. Without going too nicely into its exact value, for it differs from person to person, let us say that 40 c.c. of oxygen per minute pass through the pulmonary epithelium per mm. difference of pressure between the average pressure in the capillary and the average pressure in the alveolar air. On that showing if the oxygen consumption (total metabolic rate) were 3000 c.c. per minute, the average difference in oxygen pressure between the capillary and the alveolar air must be 75 mm. That is clearly out of the question. The average tension in the capillary must approximate much more nearly to that of the blood which leaves it than to that of the blood which enters. Therefore in the capillary of the lung, as the saturation of the arterial blood is known not to be appreciably lowered and its tension therefore known to be about 100 mm., the partial pressure of oxygen in the capillary must be over 50 mm. and therefore within 50 mm. of that of the arterial blood.

So far as we know, one of two things must happen: either there must be oxygen secretion, or there must be a large increase in the diffusion coefficient. The two most obvious means of securing the latter are distension of the lung, which makes the cells thinner, and the opening up of vascular areas, either by increasing the calibre of vessels already open, or opening up those hitherto shut. Actual observations of the diffusion coefficient during exercise have not shown a great rise. M. Krogh (1915) states that the rise as between rest and heavy work is something of the order of 30 per cent. (Subject XX, p. 286.)

More recently Ove Bøje (1933) reckons by similar methods that the diffusion coefficient in heavy work may increase by 100 per cent., although in most cases the increase is distinctly less.

The subject is treated in a different way in Professor L. J. Henderson's book *Blood*.* To quote from p. 210 of that work:

> These tables lead to the conclusion that the area of diffusing surface, which we may assume to be roughly proportional, for similar structures, to the number of physiologically active capillaries is subject to wide variation.... The coefficient of utilization of oxygen seems never to rise normally far above 60 per cent. while, in the stationary state of heavy work the oxygen consumption may be ten or even fifteenfold greater than during rest. It is evident accordingly that the diffusing capacity of the lungs may be ordinarily subject to a tenfold increase....

CAROTENE

To pass to quite a different type of phenomenon. If ten years ago I had been asked to give an example of something which, so far as I knew, had no biological significance, I might as likely as not have instanced the yellow pigment in the yolk of the egg.

The story then ran as follows: Hens eat cereals: some cereals such as maize contain the yellow pigment carotene, others such as wheat do not: from the point of view of the hen the pigmentation of the food is immaterial: carotene is soluble in fats: if therefore the hens eat maize the carotene will be deposited in the tissues in which it is most soluble: of such tissues one of the more obvious is the yolk of the egg.

The logical deduction was that if hens were fed on food from which carotene was absent (such as porridge), in time the stock of

---

* I should like to take this opportunity of testifying to the debt which I personally, and as I think students of physiology in general, owe to the author of this book.

carotene in their bodies would become exhausted and their eggs would contain colourless yolks. This was shown by Palmer and Kempster (1919) to be the case, and the observations were repeated by Plimmer, Rosedale and Raymond (1923). The next stage in this story of the production of bizarre eggs was the insertion of some other fat soluble dye in the carotene-free food, and its deposition in the yolk. Thus as was found by the above observers if the food be coloured with Sudan III the yolks become red. It is very easy to produce eggs of this character. [Personally, for what it is worth, I have never been so successful with green and blue fat soluble stains.]

Carotene was discovered and isolated from the carrot a hundred years ago by Vauguelin (1829). It is of quite wide distribution both in the animal and vegetable worlds. It colours the pollens of many flowers yellow: it exists also in leaves such as those of spinach. Occurring in juxtaposition to chlorophyll it is not visible, but when the chlorophyll disappears in autumn carotene often remains, causing the beautiful colouration which is one of the glories of nature. In animals carotene is found in the *corpus luteum* and gives to that organ its yellow colour, in biliary calculi, and, what is most interesting, in many animal and vegetable oils. The yellow colour of butter, for instance, is in part due to carotene, as is also the colour of cod-liver oil.

As the knowledge of the chemistry of this subject has increased, it has been discovered that the bodies referred to above are not quite identical.

A formula which has been given to carotene by Karrer and his collaborators (1931, 1932) is

Without going deeply into the matter it is evident that the " changes may be rung " on the above formula so that numerous bodies of slightly different composition but very similar properties may exist. Thus there turns out to be a series of carotenes in the lycopene of the tomato; the xanthophyll in leaves and the lutein of the yolk of the egg are in this category.

So much as regards the nature and occurrence of carotene; now to pass to its possible significance. It is well known that rats deprived of a certain accessory factor, or certain accessory factors jointly embraced in the term "vitamine A", do not grow as do normal rats, and that they become a prey to disease—notably a certain infection of the eyes.

To quote Javillier (1930), who gives an interesting review of carotene:

> The idea that a connection between vitamine A and the yellow carotinoid pigment was introduced into science by the American biochemist Steenbock who with several collaborators carried out this research between 1919 and 1921. The idea sprang essentially from the observation that in vegetable substances the vitamine A activity demonstrable by physiological experiment and the pigmentation of the carotinoids go hand in hand.

Subsequent research took the line of feeding young animals, hitherto deprived of vitamine A, on diets which contained carotene. The rats had been reduced to the condition in which growth had stopped and the eyes had become infected. The addition of carotene to the diet partially restored the rate of growth and completely removed the infection.

It became impossible to assert that carotene possessed no significance. But to state precisely what the significance of the yellow pigment was—that was another question. For some time the pursuit of carotene resembled that of the Will o' the Wisp. It shone out in unexpected places and ways, only to elude the worker who attempted definitely to locate it. Drummond (1919) once took the view that really pure carotene which can be obtained in crystalline form does not promote growth, while Miss Stephenson (1920) finds that butter which has been rendered colourless does so like ordinary yellow butter. Strictly, this last observation does not really deny to carotene some vitaminic action, it only denies that carotene is the sole substance in ordinary butter which is endowed with the property of promoting growth. In this connection it should be noted that a much clearer understanding now exists of the difference between vitamine A and vitamine D than at the time of Miss Stephenson's researches. In depriving animals of vitamine A, they were formerly often deprived of vitamine D in addition. It is clear that in such cases the mere addition of vitamine A to the diet could not restore normal growth.

Javillier and his pupils point out that phytol obtained from leaves is colourless and has the formula

$$\text{CH}_3 \qquad \text{CH}_3 \qquad \text{CH}_3 \qquad \text{CH}_3$$
$$\text{H}_3\text{C·CH·CH}_2\text{·CH}_2\text{·CH}_2\text{·CH·CH}_2\text{·CH}_2\text{·CH}_2\text{·CH·CH}_2\text{·CH}_2\text{·CH}_2\text{·C:CH}_2\text{·CH}_2\text{OH}$$

and that in the pure state it is powerless to promote growth, yet if even a trace of carotene be added to it—a trace so small as not appreciably to alter its iodine value—the phytol at once promotes some degree of growth. It would be tedious and superfluous to give a critical account of the work that has been carried out within the last five years on the relation of carotene to vitamine A. Moreover, to be of value such a review must needs come from a competent authority.

It seems not improbable that the crystals of so-called carotene contain three isomerides: α-carotene, β-carotene and iso-carotene (Kuhn and Lederer, 1931). According to Rosenheim and Starling (1931) the statements in the literature with regard to the optical activity of carotene are highly contradictory and they do not entirely repudiate the idea that the crystals of the mixed carotene isomerides may contain some substance which is in them but not of them and which is the substance of real biological activity. They say:

Biological tests kindly carried out by Drummond showed that our carotene fractions of high and low optical activity [presumably α and β above] produced approximately equal growth increments in rats in doses of 3 to 5 gm. The discovery of the heterogeneous nature of "carotene" and the difficulty in separating the various isomerides will delay a final answer as to their biological activity. The problem may be still more complex since, having regard to the ease with which carotene forms adsorption compounds with iodine, the possibility cannot yet be dismissed that its growth-promoting activity may be due to adsorption of another substance, chemically related or otherwise (vitamine A).

Meanwhile to the physiologist carotene research has been diverted into a very interesting channel by Moore (1931)—a channel which touches its significance rather than its chemical composition.

Moore's thesis is no less than this: that vitamine A is not carotene, but rather that carotene is converted into vitamine A in the livers of the animals into whose blood it is introduced.

I need only quote the following very striking experiment, which turns on the property of vitamine A, but not of carotene, of possessing an absorption band in the region of $325 \mu\mu$.

Rats were variously fed in Cambridge by Moore and their liver oils were sent over to Belfast for analysis by Capper (1930) with the following result:

Table XLIII

No. of rat	Sex	Diet	Band at $325 \mu\mu$
5	M.	Vitamine A free	No band
8	F.	,,	,,
9	M.	,,	,,
10	F.	,,	,,
16	F.	Vitamine A free + 0·75 mg.	Good band
17	F.	carotene per day	,,
19	M.	Vitamine A free + red palm	,,
20	F.	oil 1·5 gm.	,,
22	M.	Vitamine A free + fresh carrots *ad lib.*	,,

The conclusion must therefore be reached that carotene or some part thereof, if it prove later to be heterogeneous, behaves *in vivo* as a precursor of the vitamine.

Creatine is another substance which seemed at one time to lack a function; now it occupies a foremost place in muscular metabolism, being one of the constituents of phosphagen.

THE POSITION OF THE SPLEEN

More than one person has said to me words something like this: "The position of the spleen in the body has often struck me as being very significant. It is placed like the alimentary canal between the general arterial system and the liver, so that the blood leaving the spleen must traverse the liver before it reaches the general circulation". Is there any significance in the anatomical position of the spleen or not? Well, there may be, but it is not self-evidently so.

Granting the existence of the spleen it clearly must be some-where. If you were to place the names of the principal arteries of the body in a hat and draw for which should have the distinction of supplying the spleen, there would be quite a fair chance of the lot falling to the coeliac axis or the superior mesenteric artery.

The question then resolves itself into the following: Does either the spleen or the liver perform its functions more efficiently owing to its juxtaposition to the other?

As regards the functions of the spleen, physiology has for many years spoken with much reserve. Our knowledge of them is probably even now very incomplete. The function about which most is now known is probably that of regulating the quantity of blood in circulation. The efficient performance of this function does not seem to demand the insertion of the spleen in the portal system. It is probably desirable that it should be near the heart, so that by its sudden contraction it can materially increase the venous inflow and so "boost" the circulation rate. But for this it derives no advantage from the position which it occupies—quite the reverse. Rather might one expect the splenic vein to lead directly into the vena cava.

The only normal animal to which the spleen is a necessity, so far as I know, is the rat. I have never excised the spleen of the rat, the excision of which is usually said to be fatal. The reason, however, is a rather indirect one. The rat is very prone to infection by a form of Bartonella. This organism is kept in abeyance by the spleen; excision of the spleen allows of general and usually fatal infection of the rat by the parasite. For this purpose the spleen does not seem to derive any obvious advantage from being harnessed to the liver. I have never heard it suggested that the spleen and the liver formed a sort of double filter for Bartonella.

The next piece of definite knowledge about the spleen is the dramatic improvement caused by removal of the organ in cases of biliary jaundice. The suggestion is that the spleen exerts some sort of malign influence in this disease over the red blood corpuscles reducing their expectation of life to an alarming degree. If we go no further than what is stated in the above sentence we shall,

I imagine, not be seriously at fault for it will be observed that nothing is said:

(1) as to how the spleen spells death to the corpuscle;

(2) whether the actual act of murder is wrought in the spleen itself or alternatively by the secretion into the blood of some substance lethal to the corpuscle.

The impression is also made that this activity of the spleen is an exaggeration of its normal rôle of destroying red blood corpuscles. Here is to be sought, though perhaps not to be found, such significance as may exist in the juxtaposition of the spleen and the liver.

The idea in people's minds is derived from the association of the following facts:

(1) red blood corpuscles are seen in the spleen in various stages of degeneration;

(2) bilirubin is derived from haemoglobin;

(3) bilirubin leaves the liver in the bile.

The facts are incontrovertible although their association is problematical and their claim to form a complete picture of blood destruction would be groundless.

We must consider them in a little more detail.

It is true that corpuscles may be seen in all stages of degeneration in the spleen; but does this type of blood destruction take place on the requisite scale to account for the whole phenomenon? On that subject Peyton Rous (1923) says:

Without laboring any point it may be admitted to possess some such rôle in birds, though whether to the exclusion of other "systems" or cells, notably the cells of the hepatic parenchyma, has yet to be settled. But in the case of many, and not improbably all, mammals, some other, or at least some additional, means of blood destruction must be sought. For the number of hematophages normally encountered is obviously insufficient for the task they are supposed to accomplish. The cat has so few of them anywhere that prolonged search may be required for the discovery of a single one (178). Yet the cat possesses an active hematopoietic tissue. Is one to suppose that in this animal blood destruction does not take place *pari passu* with blood formation?

The order of quantities involved may be gathered from the following calculation—for which no other virtue is claimed than the mere portrayal of the magnitudes of the factors. The figures given are mere guesses: Suppose that a macrophage in the spleen eats one red corpuscle an hour, and is itself ten times the volume of the red corpuscle, it would eat about two and a half times its own weight of corpuscles per day. Suppose that 1 per cent. of the weight of the spleen when empty of blood consists of macrophages, this for a dog of 18 kilos with a spleen of 33 gm. would mean ·33 gm. of macrophages—which macrophages would then eat about ·8 gm. of corpuscles per day. The dog of 18 kilos would contain perhaps 1·8 litres of blood and 800 gm. of corpuscles, so that in 1000 days the whole of the corpuscles would be eaten. We do not know how long it takes on the average for the whole of the corpuscles to be destroyed; most computations, which also are known to be guesses, place the time as of a different order from 1000 days.

If the rate of destruction were underestimated ten-fold, if I had underrated any one of the factors to that extent, then the macrophagic action of the spleen could almost account for normal blood destruction. I have in the above calculation not taken into account the macrophages in the liver. Granting, however, for the sake of argument, that the macrophages in the liver and spleen are unequal to the task of destroying in the day the actual number of corpuscles which in fact are destroyed, what is the next possibility? It is that the corpuscles suffer fragmentation in the general circulation, the fragments being scavenged and "worked over" in the spleen. A "fragment" as I understand the term is a portion of a corpuscle which though broken still retains its haemoglobin and is therefore different from a "shadow". The "working over" of the fragments consists in their destruction and the degradation of the haemoglobin, presumably to the condition of bile pigment and iron.

Here I may introduce two of the facts which seem definitely to be known about the spleen, namely:

(1) that the blood of the splenic vein is richer in bile pigment than that of the general circulation; and

(2) when the spleen contracts large numbers of cells which contain haemosiderin may be seen in the splenic sinuses and splenic vein. They are not seen in the splenic artery at the same time. Haemosiderin appears to be ferric oxide or hydrate in a

colloidal form. Once in the splenic vein these cells must needs go on to the liver.

But to return to the fragments. If indeed these are all digested in the spleen, the macrophages are merely being served with the same food in the same quantities as if they had eaten the corpuscles whole. The corpuscles are, so to speak, chewed for them in the body—a process no doubt helpful to digestion, but in conception the spleen would remain as a considerable factory of bile pigments and iron.

Whether therefore the corpuscles are destroyed, in the sense of ceasing to function as corpuscles, in the spleen or elsewhere (except in the liver itself), we must count on the destruction of the haemoglobin which they had contained as located in the spleen. Perhaps not all of it, but much of it. We must count therefore on a considerable stream of bile pigments and iron as constantly leaving the spleen and being carried to the liver by the portal vein.

The liver of course provides the machinery for the excretion of the bile pigments (or such portion of them as is excreted) and for the storage of the iron. Therefore there is clearly this significance in the situation of the spleen on the portal system: certain products excreted by, or stored in the liver and produced by the spleen, reach the liver in a more concentrated form because the spleen is in the portal system, than if it were placed at random in the general circulation. The bile duct may be regarded not only as the duct of the liver but also as the duct of the spleen. In a sense therefore we may regard the spleen not as a ductless gland but as a gland (if the term "gland" is admissible) whose duct is in the liver—that conception would make of the liver and the spleen functionally a single organ.

It is at this stage that one is tempted to ask whether there is any closer connection between the two than appears on the surface.

If one regards the bile duct as a river the waters of which are coloured with bile pigment, and if one explores the hinterland in which that bile pigment is produced, it is to be found that the porphyrin-bearing cells are neither exclusively in the liver nor exclusively in the spleen, but are distributed indiscriminately in both organs. If on the map we were to colour the "macrophagic" country red, we should have a territory embracing both organs.

THE BIOLOGICAL SIGNIFICANCE OF PAIN

A person whom I will call A, aged 21 and an undergraduate, was staying at a resort much patronised by rheumatic patients where one of his relatives was undergoing a "cure". He himself had been suffering pain in the shoulder and arm which had recently become more acute and which he regarded as rheumatic but had never taken seriously; however, being in a rheumatic resort he decided to consult the doctor. Dr C informed him that he was not suffering from rheumatism but from his heart, that if he persisted in the sort of active life which he was leading he would die, but that if he took certain precautions his heart would recover. The physician to an Insurance Company gave him much the same information. The Company refused to insure A's life for more than a limited term of years and then at an abnormally high premium. A carried out Doctor C's instructions. He is now in his sixtieth year. His family is grown up. Wishing to add to his insurance after the War, the same Company granted him a policy at the normal premium.

I mention this case because it illustrates a number of points on which I shall touch, but at present I only wish to point out the fundamental one which, leaving out the intermediate stage, is this:

One of A's sons, B, when about ten years old, was taken ill with pain: a doctor staying in the house diagnosed the trouble as appendicitis. B was operated upon, only to find that his appendix "very badly needed removal". Probably the phenomenon of pain saved B's life also.

The pain associated with A's heart trouble and probably also that associated with B's appendicitis had a definite biological survival value. A simple calculation shows that had A died say at 25, the average longevity of the population of this country would have been reduced by three-quarters of a second, while even at B's present youthful age a quarter of a second has been added to the average lease of life of the Englishman and of course as A and B get old the beneficent influence of their pain will become even greater.

In addition there is the biological fact that A survived to beget his family.

Clearly, what happened to A and B is happening all around us, and although I have no statistics at my disposal to say how many lives have been saved by the sense of pain, I do not think that I run any great risk in saying that the number, especially of young lives, is sufficient to give the sense of pain a quite considerable survival value in this country.

If we look back from A, B and C to the more primitive vertebrates we see indications that pain has emerged from some primitive sensation of a more or less general kind. It is associated with the simple end-brush, which is found as the only nerve ending in the cornea and the epithelium of the penis. These two situations are not credited with the ability to feel any other sensation than that of pain. How far these endings are susceptible to chemical stimulation and how far to mechanical stimulation it is not easy to say. Nor is it easy to solve the same problem with regard to our own throats. Some of the more potent substances which have been tested have been administered as solid particles, but the effect of their contact with the mucous membrane (presumably a chemical effect) is pain.

The view is advanced by Sir Charles Sherrington (1920) that pain is from the start purposive. "Physical pain is the psychical adjunct of an imperative protective reflex." To quote Sir John Parsons (1927), "It is associated with stimuli which threaten or actually commit damage to the tissues". Indeed Parsons regards the primitive sensation of pain and the harmful trend of the stimuli which cause that sensation as being so inseparable as to warrant the following statement: "Sensory nerves which from the point of view of sensation are cutaneous pain nerves, are from the point of view of reflex action conveniently termed noci-ceptive nerves". If it be agreed that the primitive pain or its precursor evokes a protective reflex, we may ask what is the nature of this reflex so evoked?

One can imagine that it is to make the organism fight, to make it flee, or to make it immobile, any of which courses might under certain circumstances be of advantage.

The responses cited so far have been immediate and such are doubtless the only type of which the most primitive forms are capable. But at some juncture memory becomes a factor to be reckoned with and as the power of memory develops, so the memory of pain influences action, and perhaps influences it more than does the actual immediate suffering. In the familiar phrase "the burnt child dreads the fire".

Where do we stand? Having cited the cases of A and B, I passed to the consideration of the lowest forms of vertebrate life. But it will not have escaped the reader, that the cases A and B are fundamentally different from those of the lower forms subsequently considered and for this reason. The survival value of pain to a fish depends simply on the reaction of that fish to that pain. Not so with A. The survival value of the pain in question to A depended principally on the intelligence of Dr C. To say that nothing analogous could be found in the brute creation would be too strong a statement. There is of course the inherent instinct by which the mother guards her young and learns to recognise signals of distress, signals which may even be evoked by pain suffered by the offspring. Doubtless many other rather crude and primitive happenings in the animal world could be cited in which a call for help met with some effective response on the part of other members of the herd. True the part played by A was perhaps little in advance of that played by any animal in distress; but consider that played by C, consider the whole background of skill, intelligence, experience and memory which enables a doctor to say that a pain in the arm is due to an injury of some organ quite elsewhere—namely in the chest—to say what the nature of the injury may be, to say that the injury is due to overstraining the organ in question, to say that the result of such overstrain will probably be fatal and to advise the patient with beneficent results. That such a reaction of one member of the community to others should have a definite survival value to the race is surely a factor in evolution that may be regarded as quite new. Many examples in clinical medicine could be adduced in which pain amounting to agony seems to serve no useful purpose, and it is rather fashionable to adduce them. If however the whole field be reviewed, there seems little doubt that the warnings which pain furnishes lengthen life more than the actual suffering shortens it.

It is, I suppose, inevitable that having drifted within hailing distance of the subject of teleology I should incur the responsibility of taking my bearings with regard to it. At first the question arose merely with regard to the present chapter; were it conceded that every phenomenon had a significance, would that be a proof of teleology—in other words, were it clear that everything had a function, would it mean that everything had a purpose? The answer is surely "no". I have heard L. J. Henderson use some such phrase as that "the mill of evolution very soon grinds out anything which does not justify its existence". That is not teleology in the ordinary sense, it is something quite different from the conception that every part of the body had been made "just so" as part of an intelligent design. But if the detailed structure of the organism is attributed not to design but to certain laws, there still remain the questions, what is the origin of the laws? and does that origin imply some germ of teleology?

Teleology crops up again in another connection. In several places, to quote the phrase which I have used, "the stage has been set before the play commences". The reader who sympathises with my views will agree that not till the machinery for the exact regulation of the properties of the blood had been perfected up to a certain point did—I would say could—the magnificent development of man's intellectual powers take place. In the whole course of foetal development we have seen for the most part a similar sequence. The maternal factors develop not *pari passu* with the foetus but before it. The vessels of the uterus proliferate and maternal blood accumulates in them while as yet the foetus is of imponderable size. The foetal blood, that "go-between"—betwixt the mother and the foetus—is largely made before it is wanted and so forth. I agree that the phrase "the stage is set before the play commences" has a teleological ring; but a minute's consideration will show that it is not necessarily teleological. While it is the method which would be adopted by the wise housewife it is also the likely method of any form of development. The progress of foetal development clearly must be based on some sort of modified geometrical progression because the number of cell divisions at any time must be remotely connected with the number of cells present at that time to divide. The general course of foetal cell division is then fixed. Of necessity, in terms of grams, the foetus must remain minute for a long time and grow at an ever increasing rate. The growth of the uterus is not so completely a question of cell division, it is largely a question of the growth of the cells present. It is possible for it to progress from the start of pregnancy and clearly it would seem to place a more equal strain on the mother for the maternal organ to develop first and the foetal ones later. But quite apart from such special considerations it is wholly reasonable that, given the fact of development, it should take place in "stages", one stage being impossible before the previous one is established. The butterfly emerges from the chrysalis, or to quote the conversation of another of my friends—A. V. Hill—"you cannot have arithmetic until you have numbers". To say that such a conception is teleological is to commit the fallacy of saying that because all Spaniards have black hair and this person has black hair therefore this person is a Spaniard. The processes which I have described are such as might be expected were development on a teleological basis; they are quite reasonable on any evolutionary basis. The most the teleologist can claim is that, this person having black hair and not fair, teleology is not excluded.

Accidents happen in nature as elsewhere, but having regard to the above and other considerations, I range myself on the side of those who regard a phenomenon as more likely to have a significance than not. Those who think with me must shoulder the burden of discovering what the significance may be, but on our opponents rests the much heavier burden of proving the phenomenon to be an accident, if indeed it be such.

REFERENCES

BOHR, C. (1909). *Nagel's Handbuch*, **1**, 179.
BØJE, OVE (1933). *Arbeitsphysiol.* **7**, 157.
BRADFORD, J. R. and DEAN, H. P. (1894). *J. Physiol.* **16** 34.
BRODIE, T. G. and DIXON, W. E. (1904). *Ibid.* **30**, 476.
BROWN-SÉQUARD, C. E. (1872). *Lancet* and *C.R. Soc. Biol.* **4**, 180.
CAMPBELL, J. ARGYLL (1911). *Q.J. Exp. Physiol.* **4**, 1.
CAPPER, N. S. (1930). *Biochem. J.* **24**, 980.

DALY, I. DE B. and VON EULER, V. (1932). *Proc. Roy. Soc.* B, **110**, 92.
DRUMMOND, J. C. (1919). *Biochem. J.* **13**, 81.
DUNN, J. S. (1919 *a*). *J. Physiol.* **53**, iii.
— (1919 *b*). *Q.J. Med.* **13**, 49, 129.
FÜHNER, H. and STARLING, E. H. (1913). *J. Physiol.* **47**, 286.
HENDERSON, L. J. (1928). *Blood.* New Haven.
JAVILLIER, M. (1930). *Bull. Soc. Chem. Biol.* **12**, 554.
KARRER, P. and HELFENSTEIN, A. (1932). *Ann. Rev. Biochem.* **1**, 551.
KARRER, P. and MORF, R. (1931). *Helv. Chim. Acta*, **14**, 1033.
KROGH, A. (1910). *Skand. Arch.* **23**, 200.
KROGH, M. (1915). *J. Physiol.* **49**, 271.
KUHN, R. and LEDERER, E. (1931). *Naturw.* **19**, 306.
LANGLEY, J. N. (1900). Schafer's *Text-book of Physiology*, **2**, 642. London and Edinburgh.
MOORE, T. (1931). *Biochem. J.* **25**, 275.
NISIMARU, Y. (1923). *Okayama Igakkwai Zassi*, No. 397, 1.
PALMER, L. S. and KEMPSTER, H. L. (1919). *J. Biol. Chem.* **39**, 299.
PARSONS, Sir J. H. (1927). *An Introduction to the Theory of Perception*, p. 20. Cambridge.
PLIMMER, R. H. A., ROSEDALE, J. L. and RAYMOND, W. H. (1923). *Biochem. J.* **17**, 787.
PLUMIER, L. (1904). *J. Physiol. Path. gén.* **6**, 655.
ROSENHEIM, O. and STARLING, W. W. (1931). *Chem. Ind. Rev.* **9**, 443.
ROUS, PEYTON (1923). *Physiol. Rev.* **3**, 81.
SHERRINGTON, C. S. (1920). *The Integrative Action of the Nervous System*, p. 228. Yale.
STEPHENSON, M. E. (1920). *Biochem. J.* **14**, 715.
TRIBE, E. M. (1914). *J. Physiol.* **48**, 154.
VAUGUELIN, N. L. (1829). *J. Pharm. Sci. Access.* **15**, 340.
WIGGERS, C. J. (1909). *J. Pharm. Exp. Therap.* **1**, 341.

INDEX

362 INDEX

Eisenberger, 247
electrocardiogram, temperature effect
  on, 55
electrolytes in water storage, 111
"élément constant" of fats, 93
"élément variable" of fats, 93
Elvehjem, 139
  Steenbock and Hart, 139
Emotional stimuli, effect on spleen
  and voluntary muscle, 144
endocrine action, 316
end-organ, 328
Endres and Taylor, 16, 24
Endres, Matthews, Taylor and Dale,
  53
energy, fat production of, 99
Engels, 105
Enzyme action, in storage of calcium
  and glycogen, 103
enzymes, effect of temperature on, 40
epicritic sensation, 323
erectile tissue of penis, 163
ergometer, bicycle, 303
ether, anaesthesia, 147
Evans, 28, 274
  Brocklehurst and L. E. Bayliss, 273
Everest, 222
evolution, nodal heart beat in, 321
exercise, 198
  blood flow in, 200
  diffusion coefficient in, 200
  disgorgement of spleen in, 142
  effect on nodal rhythm of heart,
    320
  in anaemic anoxia, 229
  metabolic rate in, 199

fat, producer of energy, 99
  transference, 98
fats, conversion to carbohydrate, 99
  nature in plasma, 97
  oxidation hastened by haemoglobin,
    127
  storage of, 96
fatty acids in blood, 95
Feldberg, 144
Fell, Miss Honor, 101
fibres nerve, size of, 238
Fick, 190
Fischer, Hans, 139
Fischl and Kahn, 248
Fleischl, 187
Fletcher, 114, 268

Florey, 151
Florkin, 6
foetal blood, dissociation curve of, 176
  dissociation curve of haemoglobin,
    184
  hydrogen-ion concentration of, 180
foetal respiration, 14, 172
foetus, blood volume of, 148
  weight of, 154
Forbes, 232
  and Gregg, 232
Fühner and Starling, 338

galactose, 91
Gamble, 109
  and McIver, 110
Gamgee, 202
Garrey, 67
Gaskell, 267, 280
gasp, effect of foetal respiration on, 14
  effect of HCN on, 14
  in marmot, 16
Gasser, 272
  and Erlanger, 238
Gastrophilus, 73
Geiger, 182
Gelfan, 248
Gellhorn, 56
Gerard and Gelfan, 248
gill movement of fish, 64
glandular secretion and "all-or-none"
  law, 255
Glaser, 4
globin, 126
glomeruli, 264
glycerides, 95
glycogen, 91
Gorter, 138
Gowers, 187
Grab, Janssen and Rein, 156, 167
Gray, H., 141
Gray, J., 36
Greene, 221
Grollman, 220
Gunn and Underhill, 272

haematin, 74, 126
haemochromogen, 74, 126
haemocyanin, 6, 75
haemoglobin, as catalyst, 127
  buffering action of, 6, 71
  copper in, 139
  equilibrium with oxygen, 35

Printed in the United States
By Bookmasters